Content

The Importance of Catering Technology

Applied science (technology) has always been a part of catering and related courses because the factors that affect foods in their storage, cooking, preservation and quality are scientific ones. Equally in the accommodation areas cleaning science has played its part in maintaining standards.

In recent years there has been an increase in the technological areas of catering with the development of large-scale catering operations such as cook-chill and cook-freeze. These developments have resulted in parts of the catering industry following similar lines to the food manufacturing industry which reviews the process of food production from a broader viewpoint by considering such aspects as energy efficiency. This movement towards a more critical analysis of food production has resulted in four international symposia on catering equipment and design and an increase in relevant journals and trade magazines. This trend is going to continue due to the demands of the industry for more equipment that suits its particular needs.

In addition to this expansion in catering equipment the public are becoming more health conscious in the light of two recent reports produced by the National Advisory Committee on Nutrition Education (NACNE report 1983) and the Committee on Medical Aspects of Food Policy (COMO report 1984). The application of scientific knowledge is also important in the areas of food hygiene especially in the face of continuing outbreaks of food poisoning.

The need for scientific information goes further than just information to be used in the present situation. Decisions made by Government can be unhelpful to an industry and effective lobbying may become essential. For example, what affects would the catering industry experience if certain food additives were removed and could not therefore be used in certain convenience foods such as gravy mixes and dried soups? Also the Government asks various industries to comment on specific aspects of food policy. It is fitting that as the catering industry employs nearly 10% of the workforce and produced in 1984 6,000 million meals, greater than any of our European partners, that it is consulted on these issues. In *Caterer and Hotelkeeper* 17 April 1986 it was reported that the Government was interested in widespread comment on irradiated food. The reply from the catering industry, according to the article, was 'The catering industry will have a difficult task replying to the Government's report on irradiated foods by 1 July. Not least is the problem of reading it. Toxicological, kilograys, microbiological, are just a few of the phrases the reader has to take in.' If the catering industry is to make a comment and influence outside bodies then it must have the knowledge to do so.

This book is an introductory text in that the first three chapters deal with the fundamentals of chemistry, physics and biology such that the more applied areas in the remaining chapters can be built on a firmer foundation. The last chapter was designed mainly to introduce the student to catering equipment and the factors that affect performance. At the end of each chapter there are exercises which it is hoped will not only test the reader's knowledge of the text but also encourage him to seek information from other written sources and not least from the industry he hopes to serve.

1

Chemical Structure and Reactions

ATOMS

All materials are made from minute particles called *atoms*. These particles are so small that they cannot be seen with an ordinary microscope as can microorganisms such as bacteria. Oxygen atoms are 2.8×10^{-10} metres in diameter whereas bacteria are 1×10^{-6} metres, ie approximately 3000 times bigger. There are about 100 different atoms and if a material is made up from its own atoms then it is an element, eg iron and oxygen gas. It is obvious, therefore, that only about 100 elements exist, but there are many million different materials. These other materials are made up by joining together the various elements, eg water is made by joining together oxygen and hydrogen.

ELEMENTS

The elements noted in table 1.1 are the ones that you are likely to meet in an introductory food science course. The atoms are given abbreviations known as *symbols* and each atom has 'arms' so that it can join on to other atoms. The number of 'arms' an atom has is known as its *valency*. Some elements exist as separate atoms, eg aluminium metal and helium gas, but some elements have their atoms joined such as oxygen and hydrogen gases where the atoms join up in groups of two. When atoms join together the group, no matter what the size, is known as a *molecule*.

Table 1.1 *Some common elements*

Element	Symbol	Valency
Hydrogen	H	1
Oxygen	O	2
Sodium	Na	1
Chlorine	Cl	1
Calcium	Ca	2
Carbon	C	4
Nitrogen	N	1,2,3,4,5
Phosphorus	P	3,5
Sulphur	S	2,4,6
Copper	Cu	1,2
Iron	Fe	2,3
Potassium	K	1

COMPOUNDS

These are the millions of materials that are made by joining elements together. Each different compound has its own special arrangement and the only way to get to know them is to learn them, eg a water molecule is always a group of three atoms, one oxygen and two hydrogen. The reason for this is to do with the 'arms' each atom has, that is an oxygen atom with two 'arms' can join to two hydrogen atoms because they have only one arm each. A molecule therefore has completely joined up arms with a valency of zero. The water molecule is represented as H_2O.

FUNCTIONAL GROUPS

When molecules get larger, atoms tend to collect together in precise groups which function as a whole rather than as individual atoms. Spotting these functional groups can often tell you something about how the molecule will behave. The carbonate group consists of one carbon atom and three oxygen atoms arranged as follows:

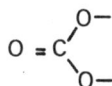

$$O = C \underset{\displaystyle O-}{\overset{\displaystyle O-}{}}$$

The group can be abbreviated to CO_3 but notice that the group still has two unfilled 'arms' giving this group a valency of two. Knowing this we can construct correctly a range of molecules containing this group, eg calcium carbonate is $CaCO_3$ but sodium carbonate is Na_2CO_3. Molecules containing the carbonate group all give off carbon dioxide gas on addition of an acid. Table 1.2 provides a list of some common functional groups with their valencies.

Table 1.2 *Functional groups and their valencies*

Name	Group	Valency
Sulphate	SO_4	2
Metabisulphite	S_2O_5	2
Nitrate	NO_3	1
Nitrite	NO_2	1
Bicarbonate (Hydrogen carbonate)	HCO_3	1
Amine	NH_2	1
Phosphate	PO_4	3
Ammonium	NH_4	1
Hydroxide	OH	1

VALENCY AND ATOMIC STRUCTURE

Atoms of course do not literally have 'arms' but this simple idea does enable us to construct quite large molecules correctly and so it continues to be used. The question is what are the 'arms'? To answer this we need to know that atoms are themselves made from even smaller particles. The structure of the atom is shown in figure 1.1.

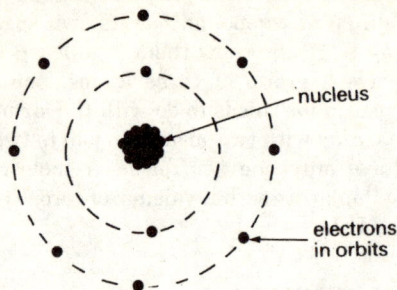

1.1   Structure of the atom

The nucleus consists of two types of particle) the *proton* which is positively charged and the *neutron* which has no charge. The electrons are about 2000 times smaller than either the proton or the neutron, negatively charged and moving around the nucleus in orbits similar to the planets going around the sun. It is obvious that the first things to touch when atoms join together is the outer orbit of electrons and it is from here that the valency comes. Before we go into valency in more detail it is perhaps a convenient place to state that as atoms are electrically neutral there must be the same number of protons as electrons. However, atoms do vary so that we can have different elements, and the variation lies in the number of protons in the nucleus, eg hydrogen has one proton, oxygen 8 protons, sodium 11 and chlorine 17. This is known as the *atomic number* whereas the number of protons plus neutrons is known as the *atomic weight*, the electrons being too small to be included.

IONIC AND COVALENT COMPOUNDS

Simple experiments to show that some compounds can conduct electricity very well, some not so well and others not at all, can be carried out using the apparatus shown in figure 1.2.

When the experiment is carried out with salty water a good electric current is observed; with carbon tetrachloride no current is seen; and with water a very small current registers. The question is how can NaCl (salt) be so different from CCl_4 (carbon tetrachloride) when chemically they look so similar? The answer is in the way the molecules are put together, that is in their valency. Sodium chloride is a combination of one sodium atom and one chlorine atom. The electrons in the sodium atom are arranged 2:8:1 and in the chlorine atom 2:8:7. It appears that all atoms strive to achieve even greater stability than they already possess. This stability is seen in the atoms of the inert gases which are so stable that they do not form molecules and have, for example, the following electron arrangements: helium 2; neon 2:8; argon 2:8:8.

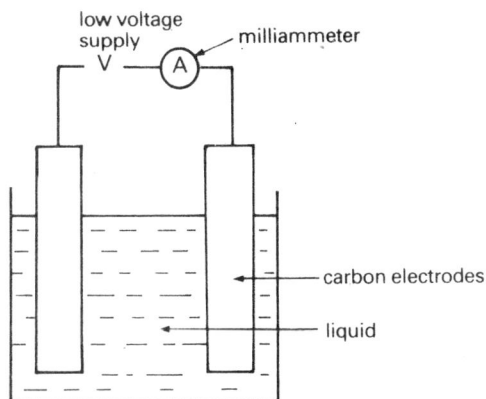

1.2   Electrical conductivity of liquids

A sodium atom tries to lose its outer electron to become like neon, and a chlorine atom tries to gain an electron to become like argon. This happens during the formation of salt by the sodium atom giving the chlorine atom an electron and by doing so the sodium becomes positively charged (it still has 11 protons) and the chlorine negatively charged. These are called *ions* and are represented as Na^+ and Cl^-. The attraction between them keeps the salt crystals together. When the salt is dissolved in water, the ions are free to move because the ionic attraction is considerably reduced and it is these ions that conduct the electricity. So the *valency* is the number of electrons an atom is prepared to give or accept. These compounds are known as *ionic compounds* and are usually solids with very high melting points and are easily dissolved in water. You can use the equipment shown in figure 1.2 to test the compounds already mentioned, for example calcium bicarbonate and sodium nitrate. On the evidence presented, carbon tetrachloride cannot produce ions but the principle of trying to achieve an inert gas structure is still operational. In carbon tetrachloride this is achieved by sharing, and the molecules are known as *covalent compounds*. The sharing is shown in figure 1.3.

The outer orbits touch and an electron of the carbon and chlorine are shared in such a way that the carbon appears to have collected four extra electrons and the chlorines one. Covalent compounds have lower melting points than ionic compounds and are usually liquids or gases.

However, it is best to mention at this point that some compounds can be ionic in one instance and covalent in another, eg hydrogen chloride gas is covalent but when it dissolves in water it becomes ionic.

We have not yet discussed the problem of water which conducts electricity slightly. The answer is in the fact that water molecules are mainly covalent but a very small number of the molecules form the ions H^+ and OH^-. This occurs with other compounds where only some of the molecules form ions, eg all acids product H^+ ions in water but some acids, such as HCl, completely ionise (strong acids) whereas others, such as citric acid, only partially ionise (weaker acids).

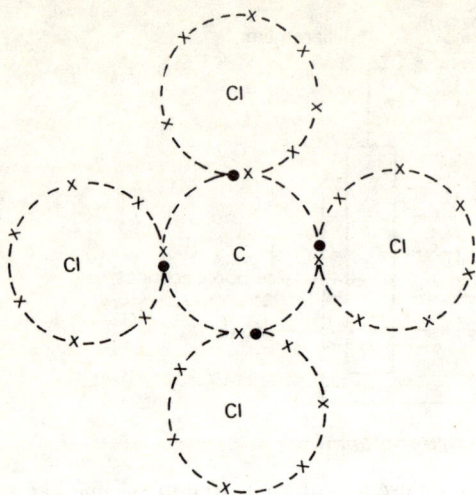

1.3 *Electron sharing in carbon tetrachloride*

CHEMICAL REACTIONS

When some materials are heated or put with other materials (reactants) a reaction takes place such that afterwards new materials have appeared in place of the old ones. When calcium bicarbonate dissolved in water is heated it disappears and new materials appear in its place. This reaction is important in removing temporary hardness from water. The reaction that is taking place is as follows:

$$Ca(HCO_3)_2 \longrightarrow CaCO_3 + H_2O + CO_2$$

Reactants   Products

The reaction shows that the products formed are calcium carbonate, water and carbon dioxide. The important points to remember about reactions are: (a) the products formed have been found by chemical investigation; (b) the products formed from the reactants must contain the same elements and (c) as no weight is lost or gained in a chemical reaction, the atom numbers must be the same at the beginning as at the end. This is known as *balancing the equation*, similar to accounting.

The chemical equations discussed have one omission in that they do not indicate whether the reaction gives out (*exothermic*) or requires (*endothermic*) heat. The type of reaction that occurs depends on the energy of the reactants compared with the energy of the products. If the products have less energy than the reactants then energy is released as heat (exothermic). The energy process can be illustrated as in figure 1.4.

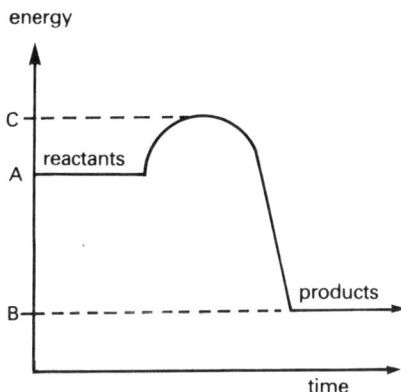

1.4   Energy changes during a chemical reaction

The energy released in the reaction is A–B and the energy that is required to get the reaction to begin C–A is known as the *activation energy*. When acids react with alkalis energy is released, eg NaOH + HCl $\longrightarrow$ NaCl + H_2O + Energy, and the activation energy is provided when the molecules of the reactants, which in solutions are moving, collide with sufficient force. If the collision does not provide the required energy then no reaction occurs. The importance of this is that reactions with high activation energies are unlikely to occur whereas those with low activation energies occur easily. Chemical reactions with high activation energies can have these lowered by putting in the reaction vessel substances known as *catalysts*, eg enzymes and some metals such as nickel. The question arises, Where does this released energy come from?, because no weight is lost during a chemical reaction. The answer is that the energy is located in the chemical bonds that hold molecules together. So the bond energies of the products in an exothermic reaction are less than the bond energies of the reactants. Exothermic reactions have been employed to release energy to heat up canned foods where stoves may not be possible, eg soldiers in war time. This reaction is shown as:

$$CaO + H_2O \longrightarrow Ca(OH)_2 \qquad 15 \text{ k.cals/mole}$$

Foods release energy so that living things can move, and in the laboratory we find out how much energy can be released from food by burning them in oxygen in a food calorimeter (figure 1.5). The reaction for the release of energy from the sugar glucose is shown as:

$$C_6H_{10}O_6 + 6O_2 \longrightarrow 6CO_2 + 6H_2O \qquad 680 \text{ k.cals/mole}$$

ATOMIC AND MOLECULAR WEIGHTS

Atoms have weight but this cannot be measured in grams because it is too small. It is, therefore, measured in atomic mass units (amu) where the hydrogen atom is given a value of 1 amu. The atomic weights of some common atoms are shown in table 1.3.

waste gases

thermometer

copper tubing

water

electric heater

food

low voltage supply

oxygen

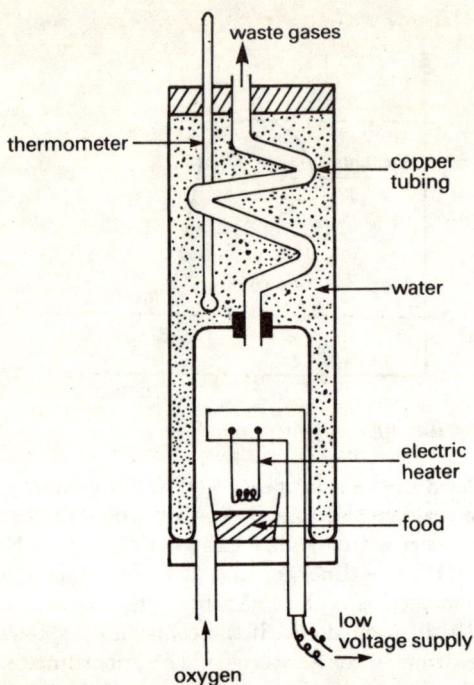

1.5   The food calorimeter

Table 1.3 *Atomic weights of some common atoms*

Atom	Weight	Atom	Weight
H	1	Cu	64
O	16	Cl	35.5
N	14	P	31
C	12	S	32
Na	23	Ca	40

As the atomic weight is the sum of protons and neutrons and these have the same weights, where does the half come from with chlorine? The answer is that there are two types of chlorine atom, ie chlorine 35 and chlorine 37, each with 17 protons but with a different number of neutrons in the ratio of 3:1 in favour of chlorine 35 which gives an average atomic weight of 35.5. These different forms of the same element are called *isotopes*. Many elements have isotopes, eg hydrogen has 3: hydrogen, deuterium and the radioactive tritium. Chemical reactions can now be expressed in weights as molecular weights are the sum of the weights of the various atoms in them, eg

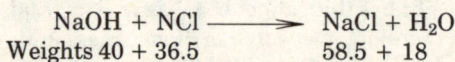

$$NaOH + NCl \longrightarrow NaCl + H_2O$$
$$\text{Weights } 40 + 36.5 \qquad\qquad 58.5 + 18$$

Therefore by balancing a reaction we can tell how many molecules are involved in a reaction and the relative weights of the reactants and products. In the case of sodium hydroxide and hydrochloric acid 40 g of NaOH would react with 36.5 g of HCl to produce 58.5 g of NaCl and 18 g of H_2O. The molecular weight in grams of a compound (the same applies to atoms and ions) is called a *mole* and so the reaction discussed could be stated as 1 mole of NaOH reacting with 1 mole of HCl to produce 1 mole of NaCl and 1 mole of H_2O. If a mole of a substance is dissolved in a total volume of 1 litre with a liquid, such as water, the strength of the solution is 1 Molar (1M).

Water molecules ionise to produce H^+ and OH^- ions with the following concentrations:

$$H_2O \longrightarrow OH^- + H^+$$

1000g $\quad$ 10^{-7} moles/litre $\quad$ 10^{-7} moles/litre

The concentration of H^+ ions $[H^+]$ is very important in relation to foods and is referred to as its pH. In the case of water the $[H^+] = 10^{-7}$ moles/litre (10^{-7} g/litre) and so according to the following formula will give water a pH = 7:

$$pH = \log \frac{1}{[H^+]} = \log \frac{1}{10^{-7}} = \log 10^7 = 7$$

The pH scale runs from 0 to 14 and so water fits into the middle. Hydrochloric acid of 1 mole/litre (1M) ionises completely to form 1 mole/litre of hydrogen ions (1g/litre):

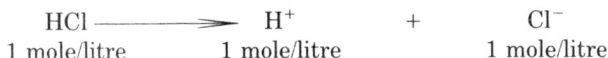

$$HCl \longrightarrow H^+ + Cl^-$$

1 mole/litre $\quad$ 1 mole/litre $\quad$ 1 mole/litre

The pH of this 1M hydrochloric acid solution using the formula already stated is:

$$pH = \log \frac{1}{1} = \log 1 = 0$$

The hydrogen ion concentration $[H^+]$ from the water in this case is ignored because it is very small compared with that from HCl. Likewise the pH of a 0.1M hydrochloric acid solution is 1 and so the stronger the acid the lower the pH. Water produces a pH = 7 because not all the water molecules ionise to form OH^- and H^+ ions. In a litre of water there are $\frac{1000}{18}$ moles (55.6) and so if this results in only 10^{-7} moles of H^+ then only 1 molecule in 556 million ionises.

When an alkali such as NaOH is added to water it produces Na^+ and OH^- ions and raises the pH above 7. The reason for this is that the $[H^+]$ multiplied by the $[OH^-]$ is a fixed number at 10^{-14}. Therefore if OH^- ions are added to water they react with the H^+ ions to produce water molecules and so lower the $[H^+]$ to less than 10^{-7} moles/litre. If a 0.1M sodium hydroxide is prepared then the pH is 13 according to the following:

$$[OH^-] \times [H^+] = 10^{-14}$$

If $[OH^-] = 0.1$ moles/litre $= 10^{-1}$ moles/litre then $(H^+) = 10^{-13}$ moles/litre. This concentration of $[H^+]$ gives a pH = 13.

Foods are divided into low acid, medium acid, acid and high acid according to their pH which can be measured using pH papers or a pH meter. The pH of food determines its preservative properties and, as table 1.4 shows, its heat processing requirements if it is to be put into a can.

Table 1.4 *Classification of foods for canning*

Classification	Foods	Heat process requirements
Low acid pH = 7	Crab meat, oysters, milk, chicken, cod, corn	High temperature processing at 121°C
pH = 6	Corned beef, peas, carrots, potatoes, asparagus	
pH = 5	Figs, tomato soup	
Medium acid pH = 4.5	Ravioli, pimentos	
Acid pH = 3.7	Tomatoes, pears, apples, peaches, apricots	Boiling water processing at 100°C
High acid pH = 3	Pickles, relish	

EXERCISES

1 Make up the correct compositions for the following compounds and give their names: calcium and oxygen; sodium and chlorine; carbon and oxygen; nitrogen and hydrogen; iron and chlorine.
2 Using the valency procedure stated construct the following molecules and find a role for the compounds in the food or accommodation area: sodium hydroxide; calcium sulphate; sodium metabisulphite; calcium bicarbonate; sodium phosphate; ammonium bicarbonate; sodium nitrite and ammonium hydroxide.
3 Draw the two forms of hydrogen chloride indicating how the valency is one in each case.
4 Complete and balance the following equations which are associated with the removal of hardness from water:

$CaSO_4 + Na_2CO_3 \longrightarrow$ calcium carbonate + sodium sulphate

$Ca(HCO_3)_2 + Na_2CO_3 \longrightarrow$ calcium carbonate + sodium bicarbonate

$Ca(HCO_3)_2 + Ca)OH)_2 \longrightarrow$ calcium carbonate + water

Ca^{2+} + ion exchange resin $- Na^+ \longrightarrow$ ion exchange resin $- Ca^{2+}$ + Na^+

2

Kinetic Theory of Matter

STATES OF MATTER

There are three states of matter: solid, liquid and gas. In a *solid* the particles, be they atoms or molecules, are stationary but vibrating. The atoms in a copper pan or the molecules in a block of ice do not fall away from each other so there must be very strong interparticle forces holding a solid together. A *liquid* and *gas* on the other hand are more fluid. The reason for this is that the particles are moving and so in a liquid the interparticle forces are very much weaker than in a solid but still sufficient to hold the liquid together. In a gas these forces do not exist as the particles are too far apart and moving at maximum speed.

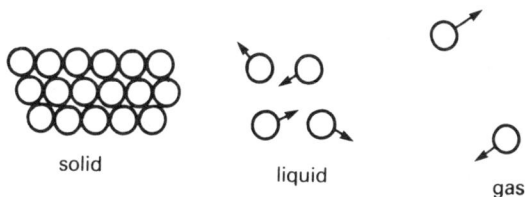

2.1 The states of matter

In the liquid state the speed of movement of the particles depends upon:

(a) the *weight* of the particle
(b) the *temperature* of the liquid
(c) the *strength* of the interparticle forces.

BOILING POINTS AND MOLECULAR WEIGHTS

If we consider two different liquids at room temperature, what experiments might we perform to evaluate the movement in each liquid? Your answer might be that the liquid with the lowest boiling point would have the most movement because the smallest temperature rise is needed to speed up the particles sufficiently to become a gas. If we take five alcohols and record their boilding points we get the results outlined in table 2.1.

Table 2.1   *Various data relating to five alcohols*

Alcohol	Formula	Molecular weight	Boiling point °C
Methyl	CH_3OH	32	65
Ethyl	C_2H_5OH	46	78
Propyl	C_3H_7OH	60	97
Butyl	C_4H_9OH	74	117
Amyl	$C_5H_{11}OH$	88	130

If the data is plotted as a graph (figure 2.2) then we can see that there is a near linear relationship between the boiling point and the molecular weight. This relationship would fit into the *kinetic theory* quite well because the particles in the various alcohols would move slower as the molecular weight increases and therefore exhibit higher boiling points.

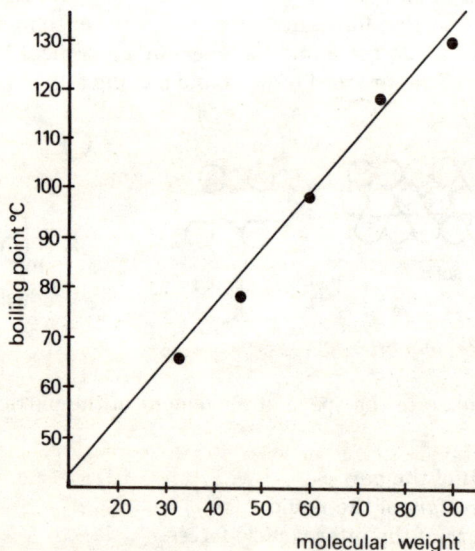

2.2   The boiling point of various alcohols

HYDROGEN AND IONIC BONDING

We have also in the graph a relationship that could be used to find the molecular weight of a substance by simply finding its boildling point. If this is true the boiling point of water, with a molecular weight of 18, would be 50°C and for sodium chloride (salt), with a molecular weight of 58.5, a boiling point of 96°C. The actual boiling points are for water 100°C and for sodium chloride 1465°C. The graph in figure 2.2 is obviously not applicable to water and salt, but why not? The answer is that the interparticle attractive forces must play an important part here and they do. Liquid salt has the particles Na^+ and Cl^-, the ions already mentioned, and the electrostatic force between the ions is very strong and reduces the particle movement.

It is worth noting at this stage that when sodium chloride solid is added to water this force is weakened to such an extent that the ions are free to move. The problem of water has yet to be resolved because it is not ionic but covalent even though a very few molecules do form the H^+ and OH^- ions (1 molecule in many millions). The answer is that in the covalent water molecule the electrons are not shared equally between the oxygen atom and the two hydrogen atoms. The electrons are pulled towards the oxygen atom making it slightly negative (S−) and leaving the hydrogen atoms slightly positive (S+). This makes the water molecules into so called *dipoles* and the small electrical attraction between the molecules, known at *hydrogen bonding*, accounts for the elevated boiling point of 100°C. This hydrogen bond is shown as a dotted line in figure 2.3.

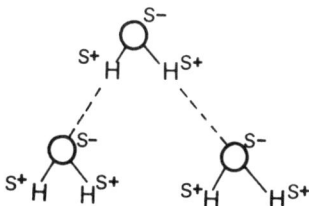

2.3 Hydrogen bonding in water

This is a convenient place to mention the fact that microwaves are able to give their energy to dipole molecules and so, as foods contain considerable amounts of water, they are used as a method of cooking.

RAISING AGENTS

The kinetic theory also explains the reason why chemical reactions that involve the collision of at least two reactants cannot take place when the reactants are solids. This property of solids is used in the preparation of baking powders because the reaction between the sodium bicarbonate and the acid is prevented until water is added and a solution is made. The reason for choosing sodium bicarbonate rather than sodium carbonate can be seen from the reactions of these two substances with hydrochloric acid, ie the sodium carbonate requires twice the amount of acid in order to produce the same amount of carbon dioxide:

$$Na_2CO_3 + 2HCl \longrightarrow 2NaCl + H_2O + CO_2$$
$$NaHCO_3 + HCl \longrightarrow NaCl + H_2O + CO_2$$

The acids in baking powders are therefore limited to solid. The question that might arise is why doesn't the baking powder on the addition of water release the carbon dioxide when the flour mixture is first made and not when you need it during the baking when the gluten (protein) in the flour can hold it like a balloon? The answer is that some powdered acids are not very soluble in cold water but dissolve as the temperature rises thus allowing collisions to take place. These acids are potassium hydrogen tartrate known as cream of tartar, calcium hydrogen phosphate and disodium dihydrogen pyrophosphate. The reaction of cream of tartar with sodium bicarbonate is as follows:

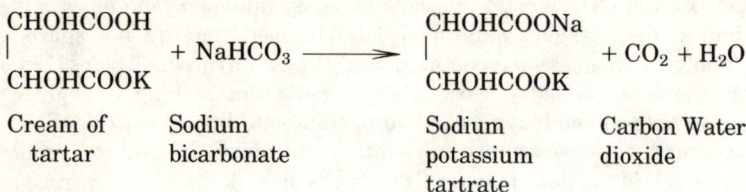

$$\begin{matrix} CHOHCOOH \\ | \\ CHOHCOOK \end{matrix} + NaHCO_3 \longrightarrow \begin{matrix} CHOHCOONa \\ | \\ CHOHCOOK \end{matrix} + CO_2 + H_2O$$

| Cream of tartar | Sodium bicarbonate | Sodium potassium tartrate | Carbon dioxide | Water |

Weights 188g + 84g

The reaction confirms the fact that the baking powder combination in this case is 100 parts of acid to 45 parts of sodium bicarbonate.

Glucono-delta-lactone has been tried as an acid because it hydrolyses to gluconic acid as the temperature rises. This reaction is not controlled by solubility but by having an activation energy large enough not to cause the reaction to occur to any extent at the lower temperature.

It is perhaps important to remember here that decomposition reactions can occur in a solid as no collision between reactants is necessary. The temperature increase vibrates the molecules until they fall apart, for example:

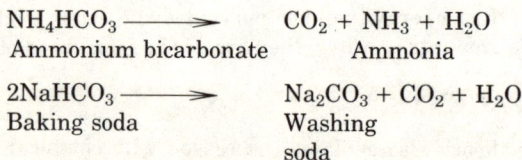

$$NH_4HCO_3 \longrightarrow CO_2 + NH_3 + H_2O$$
Ammonium bicarbonate   Ammonia

$$2NaHCO_3 \longrightarrow Na_2CO_3 + CO_2 + H_2O$$
Baking soda   Washing soda

These examples are also classed as raising agents but the products named affect the flavour and so are only used in strongly flavoured products such as ginger biscuits.

PRESSURE

Solids, liquids and gases exert pressure. In a solid the pressure exerted on a surface is expressed as its weight per unit area, eg grams per sq cm (cm^2). This pressure is produced as a result of gravity but in the case of a gas where particles are in motion the pressure is due to their density and speed. In fact the equation relating pressure (P) with density and speed is given by:

$$P = \tfrac{1}{3} \rho v^2 \tag{1}$$

where ρ = density (mass per unit volume)
v = speed (distance per unit time)

If we substitute the values of density and pressure for the gas hydrogen we find that the speed v is approximately the same as the speed of sound.

The pressure of a gas such as air is measured using a barometer (figure 2.4). The air pressure being measured can be expressed as either the height of the mercury column (h) which is 76 cm for normal air pressure at 0°C or in weight per sq cm (cm²). The air pressure (Pa) is balanced by the pressure of the mercury (PHg) in the column and so:

$$Pa = h \times \rho = 76 \times 13.6 = 1034 \text{ g/cm}^2$$
where ρ = density of mercury

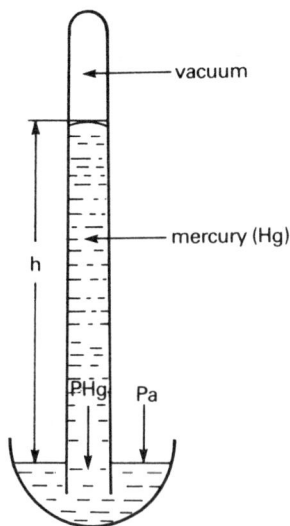

2.4   A barometer

TEMPERATURE

We usually measure temperature in degrees Fahrenheit or Centigrade (Celsius). The thermometer is used for measuring temperature and can operate due to the expansion of mercury or the electrical current produced by a thermocouple. The conversion from one scale to the other is easily carried out using the following method:

(a)   add 40;   (b)   multiply by 5/9 or 9/5;   (c)   minus 40.

EXAMPLE 1   CONVERT   100°C to °F
(a) $100 + 40 = 140$
(b) $140 \times \dfrac{9}{5} = 252$

(c) $252 - 40 = 212°F$

EXAMPLE 2   CONVERT   0°F to °C
(a) $0 + 40 = 40$
(b) $40 \times \dfrac{5}{9} = 22$

(c) $22 - 40 = -18°C$

The kinetic theory states that the temperature of a gas is proportional to its average kinetic energy. The temperature in this case is measured in degrees Kelvin ($^{\circ}$K) where 0°C $= 273^{\circ}$K (so $^{\circ}$K $= ^{\circ}$C $+ 273$).

This can be expressed as follows:

$$^{\circ}K \propto \tfrac{1}{2}mv^2 \text{ (average)} \tag{2}$$

so that as the temperature rises the speed of the particles rises as the mass remains constant. Also if the density remains constant by keeping the same volume then from equation (1) the pressure (P) increases proportionally with temperature ($^{\circ}$K). This forms the basis of the Gas Laws.

$$\text{Pressure} \times \text{Volume} \propto \text{Temperature.}$$

It is also worth noting that the average speeds of a molecule at a certain temperature depends on its mass (m). The larger the molecule the lower its speed which forms the basis of Graham's Law of Diffusion where:

$$v \propto \sqrt{\frac{1}{m}}$$

Starch and protein molecules are very large and can be considered to be stationary when dispersed in water.

PRESSURE STEAMERS AND FREEZE DRYING

The boiling point of water is only 100°C at normal air pressure. If the pressure is increased then the boiling point of water increases above 100°C.

Pressure steamers use pressures above normal air pressure so that cooking can take place in less time. The gauges show 5 lb/in.^{2}, 10 lb/in.^{2}, 15 lb/in.^{2} above air pressure such that water boils at 108°C, 115°C and 121°C respectively. Perhaps you would like to find the cooking times for pressure steamers and compare them with normal steamers.

The reduction of the boiling point of water by reducing the pressure is shown in figure 2.5. It is important to notice that it is possible to reduce the air pressure so that the boiling point meets the freezing point. This is known as the *triple point* of water and occurs at a pressure of 0.5 cm of Hg. This means that below these pressures water only exists as a solid or a gas as can be seen by the horizontal line B-----B^1 whereas above the triple point the line A-----A^1 shows that all three states of water are present depending on temperature. This phenomenon forms the basis of the freeze-drying of foods where frozen food is gently heated in a steel container having a near vacuum so that when the water is at the freezing/melting point it evaporates instead of melting (*sublimation*). Drying thus occurs at 0°C giving freeze-dried products, such as coffee, a better quality.

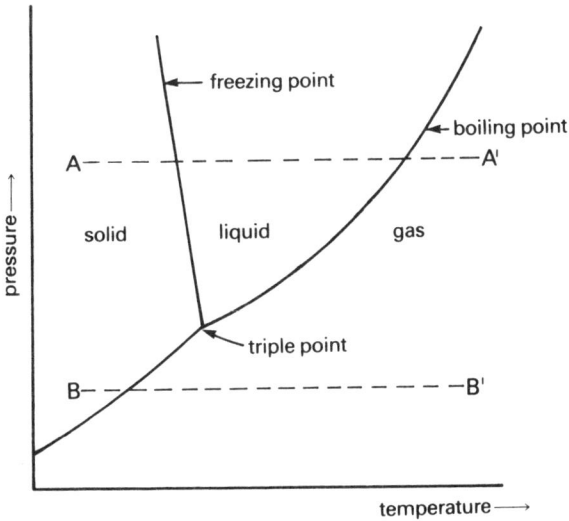

2.5   Phase diagram for water

EXERCISES

---

1   Explain why water has a higher boiling point than ethyl alcohol.
2   Describe the chemical reactions that occur in raising agents and state how the reactions are controlled.
3   Describe how the pressure of air is measured and how the control of air pressure can be used in food processing.
4   Find out the relationship between the various units used to measure pressure in steamers and compare the cooking times for normal and above normal conditions.

3

The Properties of Solutions

SOLUTIONS

When a solid, such as sugar or salt, disperses between water molecules it does so as its smallest part, ie molecules or ions, it is said to have dissolved and formed a solution. The size of the dispersed material is important because it means that the solution cannot be separated by normal filtration or by being left to stand for a time. A solution can also be created by dispersing a liquid such as ethyl alcohol in water, and a solid such as butter in a fat solvent such as carbon tetrachloride.

A solution is not the only dispersion; there are two others that will be considered in more detail in the following chapters, ie a *suspension* and a *colloid*.

DENSITY

The density of a substance is measured in weight per unit volume, ie g per ml, kg per m^3. To find the density of a regular shaped solid is fairly easy, requiring only a weighing balance and a ruler. But to find the density of a potato requires finding the volume using a displacement method. You can see from figure 3.1 that the density of a potato is an important factor when considering the type of potato used for producing chips.

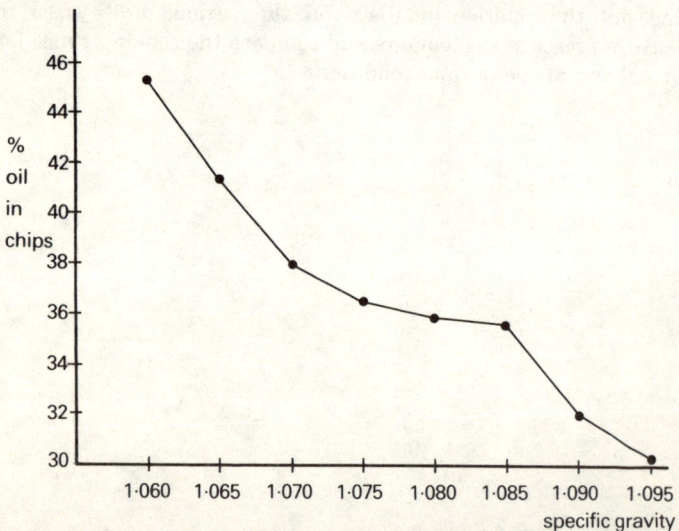

3.1   Relation of specific gravity of potatoes to oil contents of chips

The density of water under normal conditions (20°C) is taken to be 1 g/ml. If solids are dissolved in water, such as sugar or salt, the density rises but in the case of ethyl alcohol the density falls. The density of a liquid can be measured as for a solid but it is easier to use a hydrometer (figure 3.2). The density is important for measuring the strength of spirits, which are essentially a simple mixture of ethyl alcohol and water, in order to see if they have been watered down. Ethyl alcohol has a density of 0.79 g/ml and so a spirit of 40% alcohol by volume (70° proof) should have a density of 60 g water + 31.6 g ethyl alcohol/100 ml = 0.916 g/ml. Any addition of water will raise the density nearer to 1 g/ml. At this point you might ask yourself why a hydrometer cannot be used to measure the alcoholic content of beers, liquors or fortified wines such as sherry.

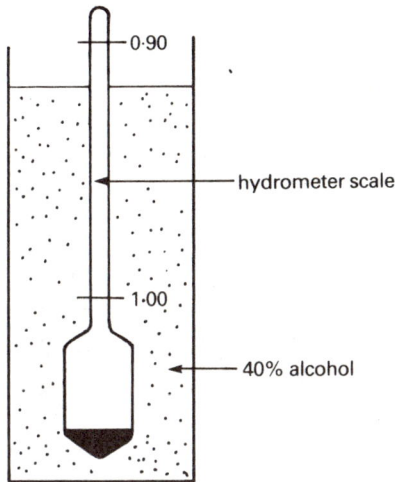

3.2   Hydrometer

It is usual today to find on canned or bottled alcoholic drinks a reference to the original specific gravity of the wort. In brewing the *wort* is the sugary solution that is converted to alcohol after the yeast has been added. The specific gravity is a measure of its density:

$$\text{Specific gravity (SP.GR)} = \frac{\text{Density of the liquid}}{\text{Density of water}}$$

As the density of water is 1 g/ml then the specific gravity of a liquid with a density of 1.2 g/ml is 1.2. The wort as it becomes changed to beer replaces sugar with alcohol and so the specific gravity falls and the hydrometer sinks deeper into the liquid. Today the density of a liquid can be measured more accurately and quickly by using the piece of equipment shown in figure 3.3.

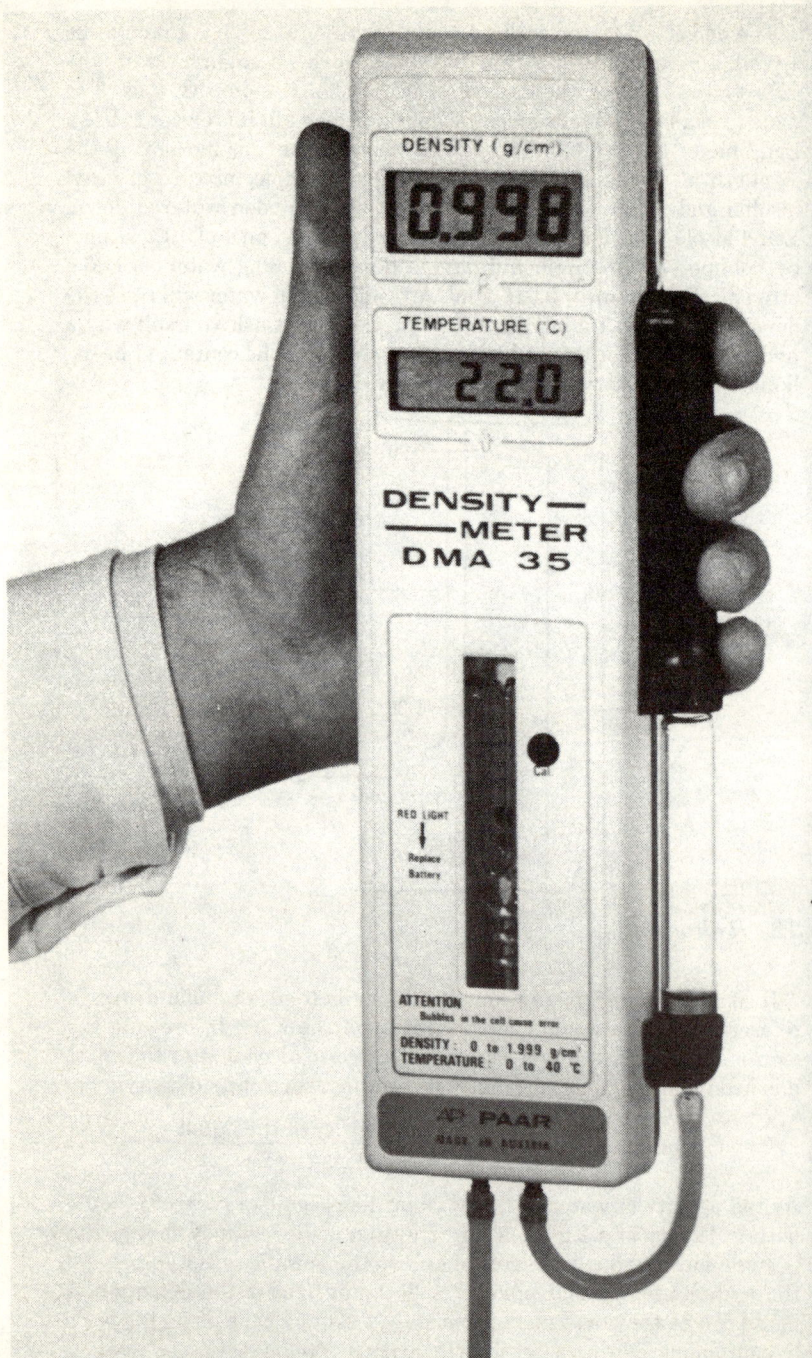

3.3   Digital density meter

BOILING AND FREEZING POINTS

When substances such as salt or sugar are dissolved in water the boiling point of the solution is raised above the boiling point of water (100°C) and the freezing point lowered below the freezing point of water (0°C). These factors account for the temperatures that are possible in sugar boiling where evaporation of the water causes the solution to rise further in temperature as the solution gets stronger.

The depression of the freezing point by dissolved solids, such as sugars and amino acids, accounts for the fact that food does not freeze at 0°C but at a slightly lower temperature enabling foods that are stored at 0°C (chilled) to keep water in its liquid state. Strong sugar and salt solutions can also be used to freeze foods as they remain liquids down to about −24°C. It is also possible to see if milk has been watered down by measuring its freezing point.

DISTILLATION

Distillation is a technique to separate substances that have different boiling points. In the field of food science one of its main applications is in the production of spirits, such as brandy and whisky, from fermented vegetable sources. In the production of wine the alcoholic content reaches a maximum of 18% by volume and so to obtain a spirit of some 40% by volume, the alcohol must be separated from the remainder of the wine. It is possible to separate the alcohol from the water by lowering the temperature as water freezes at 0°C and ethyl alcohol at −133°C, however, wine contains other more poisonous alcohols besides ethyl alcohol, eg propyl, butyl, known as the *fusel oils*, as well as acetaldehyde, which must be removed.

The simple still known as the *pot still* is shown in figure 3.4. It consists of a boiler which contains the fermented liquor which is heated and, as the boiling point of ethyl alcohol is 78°C, the vapour condensing in the coiled pipe is richer in alcohol than in the boiler. This method is still used for making brandy and whisky where the fermented liquor is distilled twice. On the first distillation all the alcohol is collected but on the second the first vapours containing acetaldehyde (CH_3CHO), with a boiling point of 28°C, and the final vapours containing the fusel oils are discarded. It can be seen that the process is laborious and the separation of the poisonous acetaldehyde (*heads*) and the fusel oils (*tails*) from the ethyl alcohol (*good heart*) are not always complete.

In 1830 Aeneas Coffey invented the continuous still which is based upon fractional distillation where evaporation and condensation occur more than once in the same piece of equipment. A graph to show the effects of fractional distillation is shown in figure 3.5 where the liquid line shows the boiling point of an ethyl alcohol/water mixture and the vapour line shows the equilibrium composition of the vapour resulting from that mixture. A mixture with a composition A will produce an equilibrium vapour of composition B which after condensation will, if reheated, produce a vapour of composition C. The vapour and the subsequent liquid will therefore become progressively richer in ethyl alcohol until a maximum concentration of 96% ethyl alcohol is achieved.

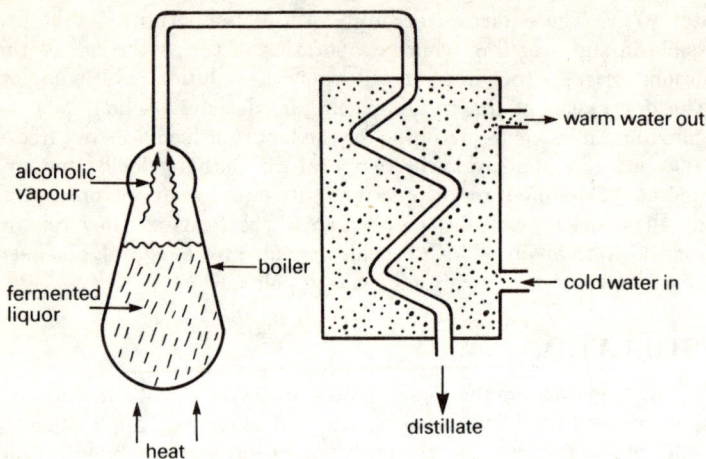

3.4   Pot still

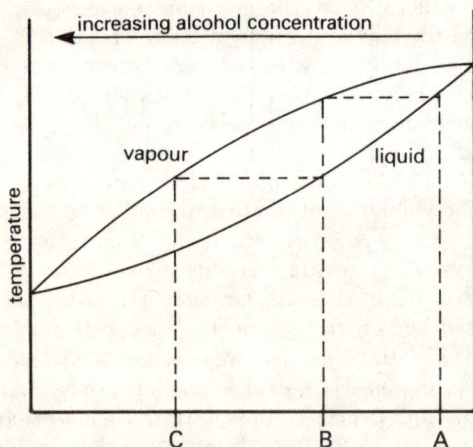

3.5   A boiling point equilibrium diagram

The Coffey still as shown in figure 3.6, shows the alcoholic vapours entering the rectifier condensing and revaporising on the perforated plates. The rectifier is in fact a fractionating column and at the position in the column of 78°C the almost pure ethyl alcohol is collected using a solid plate. The less volatile fusel oils will not reach this part of the column and so will be removed as tails and the more volatile acetaldehyde will leave the top of the column as the heads. The continuous still results in a much better separation of the good heart from the poisonous heads and tails.

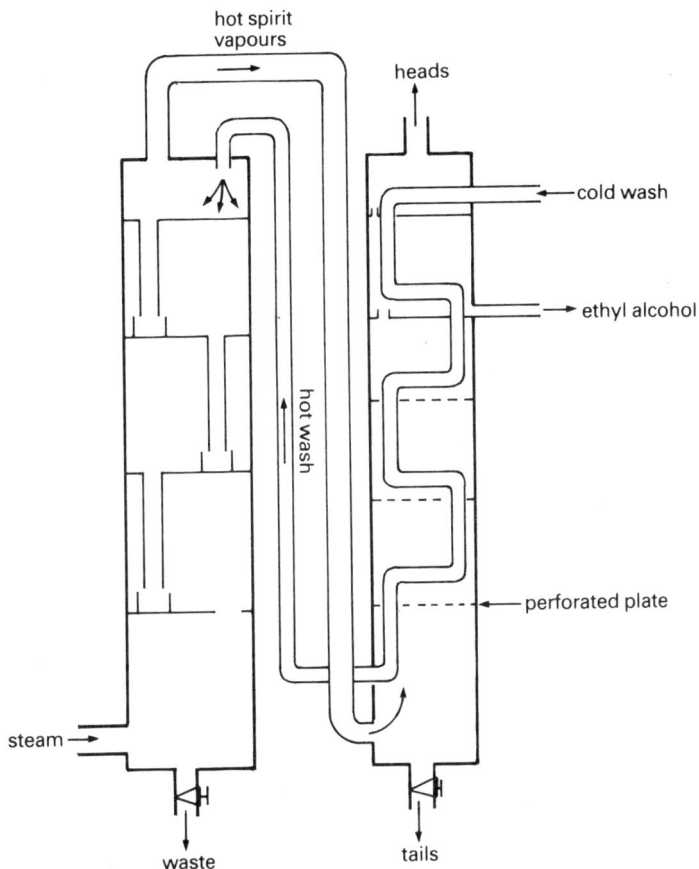

3.6   Coffey still

It is possible to carry out distillation at a lower temperature by reducing the air pressure above the liquid using a vacuum pump. This method is employed when the separation of heat-sensitive materials is required. This is known as *vacuum distillation*.

RELATIVE HUMIDITY AND OSMOTIC PRESSURE

Relative humidity is a measure of the water vapour content of air.

$$\text{Relative humidity } (\%) = \frac{\text{Weight of water vapour in air}}{\text{Weight of water vapour in saturated air under identical conditions}} \times 100$$

It is measured using a hygrometer and the one shown in figure 3.7 is known as a *wet and dry bulb hygrometer*. The principle behind this hygrometer is based upon the fact that the drier the air the faster water will evaporate from the moist cloth around the wet bulb. Evaporation causes cooling of the water remaining in the cloth because, as explained by the kinetic theory, only the faster moving water molecules will escape as a vapour, leaving behind water molecules with a lower average kinetic energy. The wet bulb thermometer will therefore show a lower reading than the dry bulb and the greater the difference (D) between them the drier the air the lower the relative humidity. To convert the temperature readings into relative humidity you need to refer to a table of relative humidities (table 3.1).

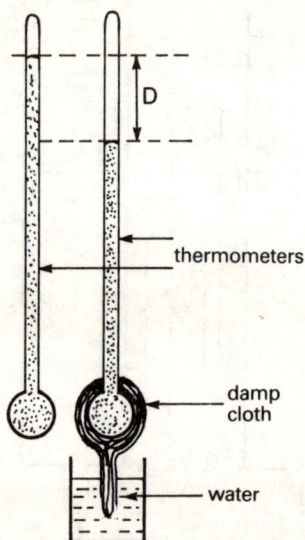

3.7   Hygrometer

Table 3.1   *Relative humidities*

Dry bulb °C	Wet bulb depression (D) °C					
	1	2	3	4	5	6
0	82	65	48	31		
2	84	68	52	37		
4	85	70	56	42		
6	86	73	60	47	35	23
8	87	75	63	51	40	29
10	88	76	65	54	44	34
15	90	80	71	61	52	44
20	91	83	74	66	59	51
25	92	84	77	70	63	57

Relative humidity is very important in regard to the working conditions in areas where the production of steam can raise it causing the natural cooling of the body by evaporation of sweat to be impaired. Relative humidity is also important in the cold storage of perishable foodstuffs where, if the humidity is too low, the food dries out but if it is too high, it encourages the growth of moulds and bacteria. Most fruits are best kept at humidities of about 90%, leafy vegetables even higher at 95%, whereas onions maintain their quality at a lower humidity of 70%.

If we consider a situation where a food is kept in an enclosed container with a small air space above it, the humidity in the air reaches a fixed value known as the *equilibrium relative humidity* (ERH).

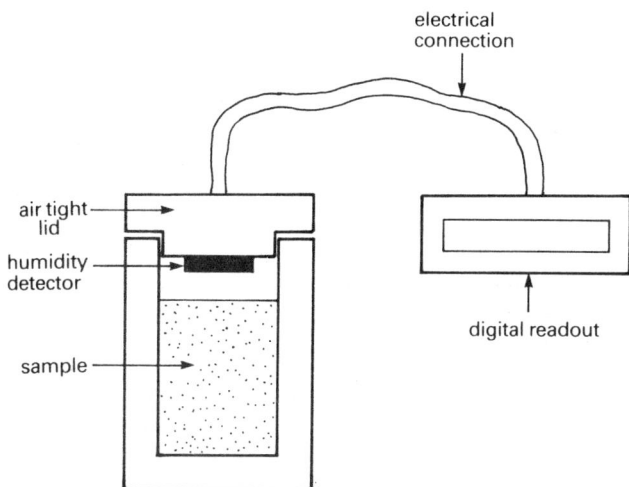

3.8   Measuring water activities

At this stage the molecules of water evaporating from the liquid surface equal the number returning to the liquid. The equipment shown in figure 3.8 is a modern method of determining the ERH of materials placed in an enclosed container. If we place water in the container and wait for equilibrium to occur, the digital read-out will give a value of 100. However, if we place into the container a solution, the ERH will fall depending on the concentration of the solution because the molecules of water are restricted in part from escaping as a vapour due to collisions with the other non-evaporating molecules at the surface.

$$ERH \propto \frac{1}{\text{concentration of solution}}$$

The more concentration the solution the lower the ERH. In the biological world the ERH is known as the *water activity* (a_w) of the solution and, as you will see, is also related to the osmotic pressure.

The a_w of a solution of, say, sucrose in water can be found using Raoult's Law where:

$$a_w = \gamma \cdot \frac{N(H_2O)}{N(H_2O) + N(sugar)}$$

$N(H_2O)$ = moles of water
$N(sugar)$ = moles of dissolves sugar (solute)
γ = activity coefficient which equals 1 for an ideal solute.

The calculated a_w for a solution of 50 g of sucrose (MW = 342) dissolved in 100 g of water is:

$$a_w = \frac{\dfrac{100}{18}}{\dfrac{100}{18} + \dfrac{50}{342}} = \frac{5.55}{5.55 + 0.146} = 0.975$$

Note that the a_w is not converted to a % by multiplying by 100 as in the case of the ERH.

Experiments carried out using the instrument described show that the observed results given an a_w less than the calculated result and the more concentrated the solution the greater the deviation becomes. The reason for this is that the solutions used are not ideal solutions where the attractions between molecules are the same. The results in figure 3.9 show the a_w's obtained experimentally at 20°C using solutions of sucrose, fructose, glycerol and salt. It can be seen that, weight for weight, fructose, glycerol and salt reduce the a_w better than sucrose. These are known as *humectants*. The reduction of the a_w in products such as bread, cakes and pizza bases by the addition of sugar, etc, extends the shelf life by reducing microbial growth. It should be remembered that substances such as salt ionise to Na^+ and Cl^- and so the effective molecular weight in this case is not 58.5 but 29.25. The a_w is therefore reduced more effectively, weight for weight, by low molecular weight ionic solutes but the use of such solutes is determined by their effect on the flavour of the commodity, hence the limited use of salt.

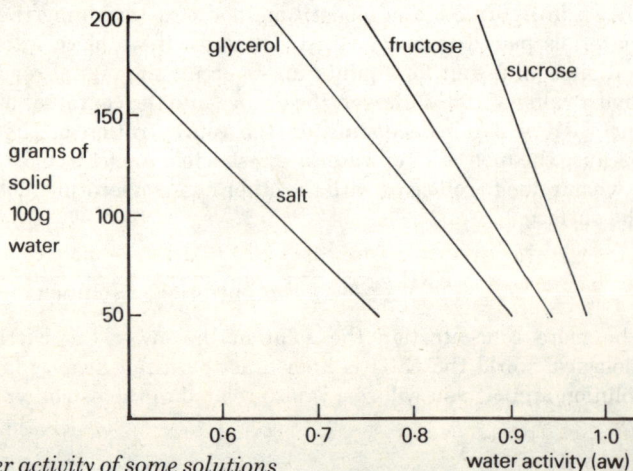

3.9 Water activity of some solutions

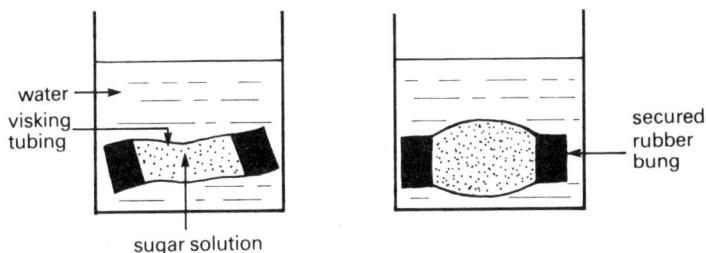

sugar solution

3.10   Osmosis

OSMOSIS

When a solution of sugar is separated from water by a membrane that only allows the passage of water molecules, the water diffuses into the sugar solution causing an increase in volume as shown in figure 3.10. The movement of water follows the direction of the highest concentration to the lowest concentration, ie from a weak solution to a more concentrated solution. The movement of water creates a pressure inside the visking tubing 'cell' and is known as the *osmotic pressure*. The osmotic pressure (OP) is proportional to the concentration (c) and the absolute temperature (T):

$$OP \quad \propto \quad cT$$

A mole of solute dissolved in a litre of water at 0°C exerts an osmotic pressure of 22.4 atmospheres. A mole of any substance contains the same number of molecules and is known as *Avogadro's Number* (approximately 6×10^{23}). It is therefore important to note that the osmotic pressure is dependent on the number of molecules present as for a_w and so the lower molecular weight substances, weight for weight, give the higher osmotic pressures.

If the experiment in figure 3.10 is repeated with the same concentration of salt or starch the visking tubing cell shows little or no increase in volume – can you suggest a reason for this?

Osmosis can occur across any membrane, and potato slices can be shown to swell or shrink (change in weight) by immersion in water or increasingly stronger solutions. In this case salt solutions will cause osmosis showing the difference between artificial and biological membranes.

A relationship between a_w and osmotic pressure exists because they are both dependent on the concentration of the solution at the same temperature but a_w decreases as the concentration increases whereas OP increases.

$$a_w \quad \propto \quad \frac{1}{OP}$$

When water is withdrawn from microbial cells by osmosis, because they are surrounded by a stronger solution they cannot grow and so the foods are preserved. From a practical point of view finding the a_w

of a solution or food material using the apparatus described in figure 3.8 is far easier than finding its osmotic pressure.

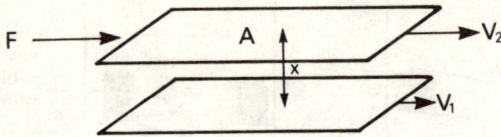

3.11   Frictional force, created by moving layers in a liquid

VISCOSITY

Viscosity is the 'thickness' or consistency of a liquid. The viscosity increases as concentration increases as can be seen with sugar syrups but it is well known that starch can thicken water at a far lower concentration than sugar. It requires a 67% sugar concentration to form a normal syrup but a thick sauce can be achieved with only 6% starch.

All liquids have a viscosity and it is associated with the frictional forces that exist when layers within the liquid flow at different velocities (figure 3.11) as when pouring wine from a bottle. The coefficient of viscosity (n) measured in poise is given by the following formula:

$$n = \frac{\text{Shear force}}{\text{Rate of shear}}$$

The *shear force* is the force per unit area (F/A) required to produce the flow or rate of shear. The rate of shear is measured by the velocity gradient between the moving layers ($\frac{V_2 - V_1}{x}$).

This formula was devised by Issac Newton and he assumed that at a given temperature the viscosity was independent of the rate of shear, ie if you want to move a liquid twice as fast it will require twice as much force.

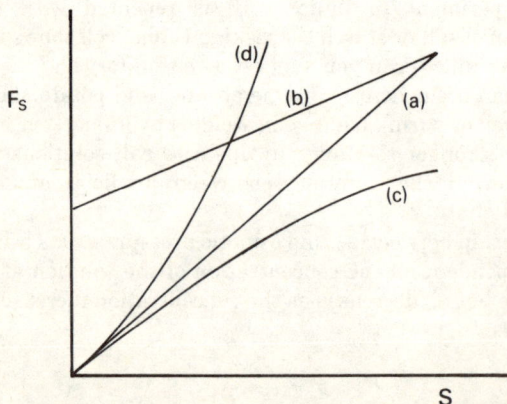

3.12   Relationships between shear force (Fs) and rate of shear (S) in various liquids

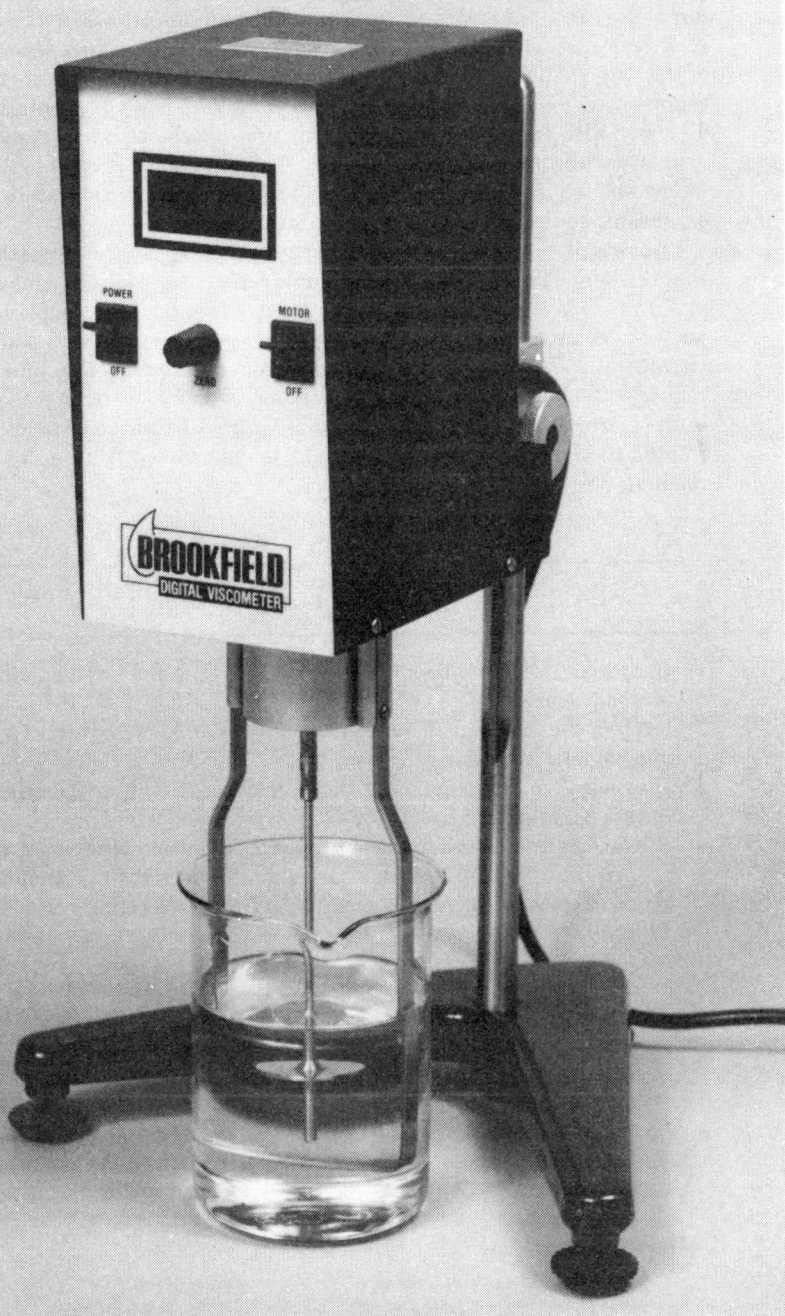

3.13   *Digital viscometer*

The graphs in figure 3.12 show the relationship between the shear force (F_s) and the rate of shear (S). Graph (a) shows a typical Newtonian liquid where the viscosity remains constant, independent of the rate of shear, eg water and thin oils; (b) shows a material that requires an initial push before it will flow; (c) shows a decreasing viscosity with increasing rate of shear and (d) shows an increasing viscosity with increasing rate of shear. The liquids (b), (c) and (d) are known as *non-Newtonian liquids* and most liquid foods such as emulsions, soups and sauces fit into this category.

Viscosity is a quality factor in food preparation as it is important that thickened liquids have an acceptable consistency. This is achieved in a normal catering operation using the experience of the chef but in large-scale production this quality factor is determined using a viscometer. Viscometers have ranged from measuring the time it takes for the liquid to flow out of a standard cup with a hole in its base to the more sophisticated type shown in figure 3.13 where a spindle is rotated in the liquid and the viscosity is determined from a digital readout.

EXERCISES

---

1 Describe the scientific processes involved in the production of alcoholic spirits.

2 Explain the relationship between relative humidity, water activity and osmotic pressure.

3 How does temperature affect the properties of solutions discussed in this chapter?

4 A sauce can thicken after freezing and reheating. How could you ensure that the consistency after reheating was the one that you required without having to modify it at this later date?

4

Food Molecules

SIMPLE ORGANIC SUBSTANCES

Food molecules are based on the chemistry of the carbon atom. The carbon atom has a valency of 4 and these 'arms' are arranged over the surface giving the filled-in shape of a tetrahedron as shown in figure 4.1. The importance of this is to convey the fact that food molecules are three dimensional and not flat as they look on these pages.

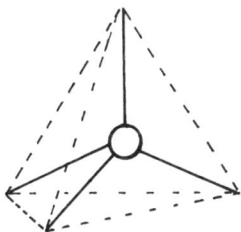

4.1   The tetrahedral carbon atom

The simplest organic molecules containing carbon are the hydrocarbons which consist of carbon and hydrogen. The smallest hydrocarbon molecule is methane (CH_4) but as carbon atoms join with each other larger molecules are formed such as propane (C_3H_8), butane (C_4H_{10}) and octane (C_8H_{18}). These chain molecules have single carbon to carbon bonds which are arranged in a zigzag pattern and have a general formula of C_nH_{2n+2}. However, there are other hydrocarbons that do not fit into this formula, for example the plant hormone ethylene, responsible for the ripening of fruit, has a formula C_2H_4. The structure of ethylene is shown to have a double bond between its carbon atoms placing it into the category of 'unsaturated carbon compounds'. Other molecules such as cyclohexane and benzene have a cyclic or ring structure showing once again the properties of saturation and unsaturation. These and other molecules are shown in figure 4.2 along with their usual abbreviated forms.

In this introduction other compounds besides hydrocarbons need to be mentioned. We have already met some alcohols where an OH group replaces one of the hydrogens of a hydrocarbon giving rise to the series methyl, ethyl, propyl, etc alcohols. The OH group is referred to as a functional group because it behaves similarly, wherever it is found, giving the alcohols certain chemical properties. The alcohols also

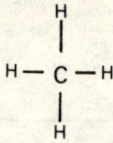
methane

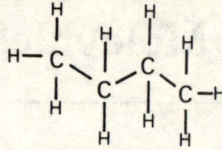
butane

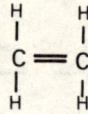
ethylene

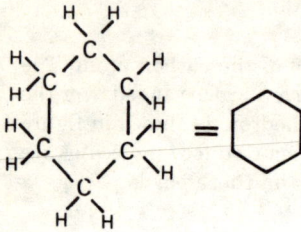
cyclohexane

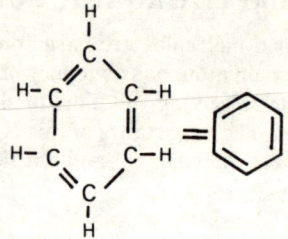
benzene

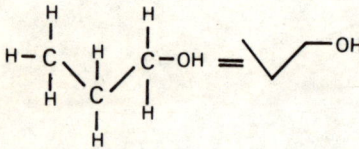
n-propanol

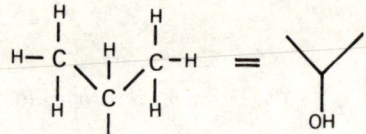
isopranol

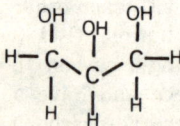
glycerol

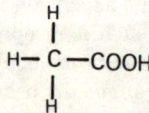
acetic acid

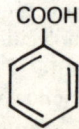
benzoic acid

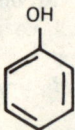
phenol

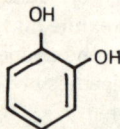
catechol

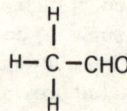
acetaldehyde

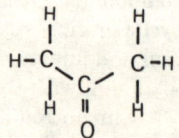
acetone

4.2   Some organic molecules

introduce the topic of isomerism where a molecule has more than one form because of the different spacial arrangement as shown with propyl alcohol (figure 4.2). Alcohols with more than one OH group are present in foods, more notably glycerol, found in fats. The acid functional group is COOH (carboxyl) with examples of acetic acid (vinegar) and benzoic acid (a preservative) shown in figure 4.2. When the OH group is attached to a benzene ring it becomes a phenol and these also show isomerism when more than one OH group is present. Phenols form the basis of disinfectants and are involved in the browning reactions that occur when apples and potatoes are cut and exposed to the air. The functional groups that constitute the aldehydes and ketones are CHO and CO respectively with acetaldehyde, already referred to as the poisonous heads during the distillation of wine, and acetone a cleaning agent, as suitable examples.

LIPIDS

The term *lipid* is used to cover fats and oils proper and fat-like materials that have similar solubilities, ie they are almost insoluble in water but soluble in organic solvents such as benzene, ether, chloroform or chloroform-methanol mixtures. Molecules that are soluble in water are referred to as *polar* materials and have similarities with water in that they contain OH or ionic groups, whereas materials insoluble in water are *non-polar*. Fats and oils proper are esters of the trihydric alcohol glycerol and unbranched monocarboxylic acids known as the *fatty acids*. The resulting molecule is known as a *triglyceride* and, as there are many different fatty acids naturally occurring, fats and oils always contain mixtures of numerous triglycerides. The structural formulae of a triglyceride and various fatty acids are shown in figure 4.3. Fatty acids can be saturated, monounsaturated and polyunsaturated depending on the number of double carbon to carbon bonds in the fatty acid chain. The configuration about the double bond is normal cis but the trans form does exist.

The difference in the fatty acid chain in having cis as opposed to trans is that *trans* gives a straight chain whereas *cis* gives a curved configuration as shown in figure 4.4. The fatty acids are often abbreviated to a total carbon number and the number of double bonds, such that palmitic acid would be $C_{16:0}$ and linolenic acid $C_{18:3}$.

When fats and oils are heated they eventually reach a temperature where they begin to smoke. This temperature, not surprisingly, is called the *smoke point*. For deep fat frying where the oil is reused it is important that the smoke point be above the frying temperature which is usually between 160°C and 180°C. Table 4.1 shows the smoke points for fresh samples of some common fats and oils although it must be pointed out that these figures can vary depending on the variability of the source. It can be seen that lard and groundnut oil are the most suitable but, because of the considerations to be discussed later, vegetable oils are usually used.

palmitic acid

oleic acid

linoleic acid

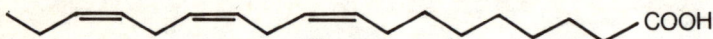

linolenic acid

arachidonic acid

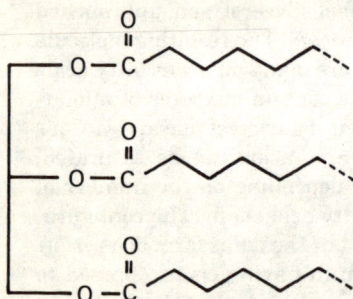

triglyceride   cholesterol

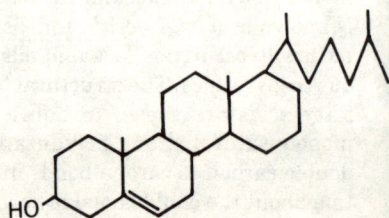

4.3   Molecules associated with lipids

Table 4.1 *Smoke points and flash points of some common lipids*

Lipid	Smoke point °C	Flash point °C
Ground nut oil	240	330
Lard	220	320
Palm oil	220	320
Olive oil	170	
Beef dripping	165	
Coconut oil	140	
Oleic acid	105	

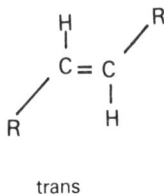
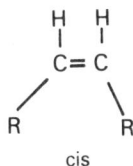

trans   cis

trans   di-cis   mono-cis

4.4   *Cis and trans fatty acids*

The smoke point of an oil during frying falls as the oil is used because the oil molecules undergo deterioration. The changes that occur also give 'off-flavours' in the oil which when absorbed by the food, can be 10% or more, make it rancid and therefore unpalatable. In a recent study it was found that the absorption of fat in chips increased as the size of chip decreased, doubled if the chips were fried in two sessions rather than a single fry and absorbed more fat if the product was purchased frozen. This deterioration occurs because of hydrolysis, pyrolysis and oxidation of the triglyceride molecules. Hydrolysis occurs in the presence of water and results in the triglyceride breaking down to free fatty acids and glycerol. The water comes from the food being fried and in the presence of such high temperatures the longer chain fatty acids are broken down to smaller chain fatty acids which are sour to the taste and the glycerol to an acrid smelling substance called *acrolein*.

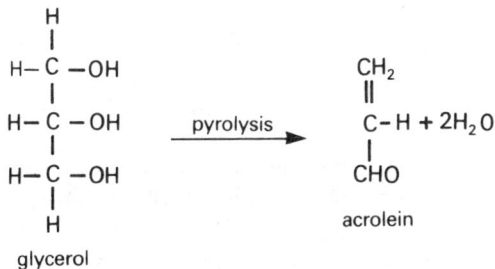

glycerol

Oxidation in fats and oils occurs mainly because of the unsaturated bonds in the fatty acid chains. The sequence of events results in the formation of radicals which react with oxygen to form hydroperoxides. The hydroperoxides are odourless intermediates which break down to smaller molecules which produce the rancidity. The oxidative sequence has three stages (1) *initiation* which involves the formation of free radicals (2) *propagation* which involves the free radical chain reactions and (3) *termination* which involves the formation of non-radical products. The initiation is the removal of a hydrogen atom from the carbon atom adjacent to a double bond. The propagation is the reaction with oxygen that leads to an unstable peroxide free radical (ROO$^\bullet$) which produces hydroperoxides (ROOH) and further free radicals (R$^\bullet$) for the process to continue. This self-propagating chain can be terminated by radicals combining to give products that do not need the propagating reactions. The sequence can be represented as follows:

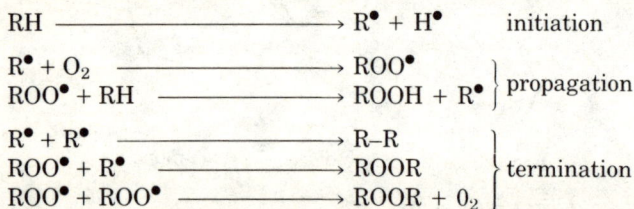

$$RH \longrightarrow R^\bullet + H^\bullet \qquad \text{initiation}$$

$$\left.\begin{array}{l} R^\bullet + O_2 \longrightarrow ROO^\bullet \\ ROO^\bullet + RH \longrightarrow ROOH + R^\bullet \end{array}\right\} \text{propagation}$$

$$\left.\begin{array}{l} R^\bullet + R^\bullet \longrightarrow R\text{--}R \\ ROO^\bullet + R^\bullet \longrightarrow ROOR \\ ROO^\bullet + ROO^\bullet \longrightarrow ROOR + O_2 \end{array}\right\} \text{termination}$$

The list of secondary products from the breakdown of hydroperoxides (ROOH) include hydrocarbons, alcohols, ketones, aldehydes and acids making the fat more polar in nature. Other more visible changes occur in that the fat becomes darker in colour, easily foams and increases in viscosity due to the formation of polymers at the termination stage.

The factors that increase the deterioration in fats and oils whilst being used in deep fat frying are increases in temperature, water, oxygen, metals such as iron and copper, and charred food debris. For the caterer the extension of the frying-life of fats and oils is important from an economic (representing 30% of the cost of a portion of chips) and food quality point of view and to meet this objective the following points will be of help:

(1)   Check oil temperature and thermostat performance.
(2)   Reduce oil temperature if fryer is inactive for any length of time and cover oil with a floating aluminium cover when fryer is not in use.
(3)   Use stainless steel frying baskets.
(4)   Avoid introducing foreign matter into frying, eg salt, cleaning materials especially alkalis, water on chips, and remove food debris.
(5)   Check regularly for signs of oil deterioration such as excess foaming, smoking, objectionable smell and bitterness of taste.

In recent years equipment designers have tried to include in their products easy methods of extending frying-life of oils. These are (1) The use of temperature probes (computer linked) to improve frying temperature control such that operators do not need to be dependent on fryer thermostats which are notoriously unreliable; (2) the cold zone fryer which was designed to collect food debris, and the more modern fryers which have built-in filter systems especially where they are employed in large-scale frying of breaded products such as chicken. There has also been a development to remove some of the polar deterioration products by the addition of powders to which these polar products adher and can be subsequently removed by a fixed or mobile filtering unit. The manufacturers of such products claim that the frying life can be extended by approximately four times over simple filtering through a filter paper.

The use of such large volumes of oil in fast food operations has led to a need for a check on oil quality so that they can be discarded at the right time. The chemical methods that are usually employed are:

(1) Measuring the % acidity in the oil where the amount of acid is found by adding a known amount of alkali to neutralise it. Fats for sale must have a % acidity below 1.5.

(2) Measuring the peroxide value which measures the hydroperoxide content of the oil. Rancidity is detected with values between 10 and 20 whereas fresh products have a value less than 5.

(3) Measuring the % of polar material in the oil by passing a sample through a column containing an absorbent which separates the polar material from the non-polar so that they can be individually weighed.

(4) Measuring the amount of malonaldehyde $CH_2(CHO)_2$, a break-down product of oxidation, by reacting it with thiobarbituric acid (TBA) and measuring the optical density at 532 nm with a spectrophotometer. Malonaldehyde formation from linolenic acid $C_{18:3}$ is shown in figure 4.5.

In Germany, frying oils must be discarded when the non-polar portion reaches 27%. The quest to find an easy-to-use rancidity test for caterers has had limited success with the sale of test kits and the introduction of electrical conductivity meters. It should be mentioned at this point that rancidity can be produced in fats by irradiation. The irradiation given either as a gamma ray from radioactive substances such as cobalt 60 or from high energy electrons ionises water according to the following sequence:

$$H_2O \rightsquigarrow H_2O^+ + e^-_{aq}$$

$$H_2O^+ + H_2O \longrightarrow {}^{\bullet}OH + H_3O^+$$

$$^{\bullet}OH + RH \longrightarrow R^{\bullet} + H_2O$$

The $^{\bullet}OH$ radicals produced by the radiolysis of water remove hydrogen atoms from the fatty acid chain to begin the reactions already discussed.

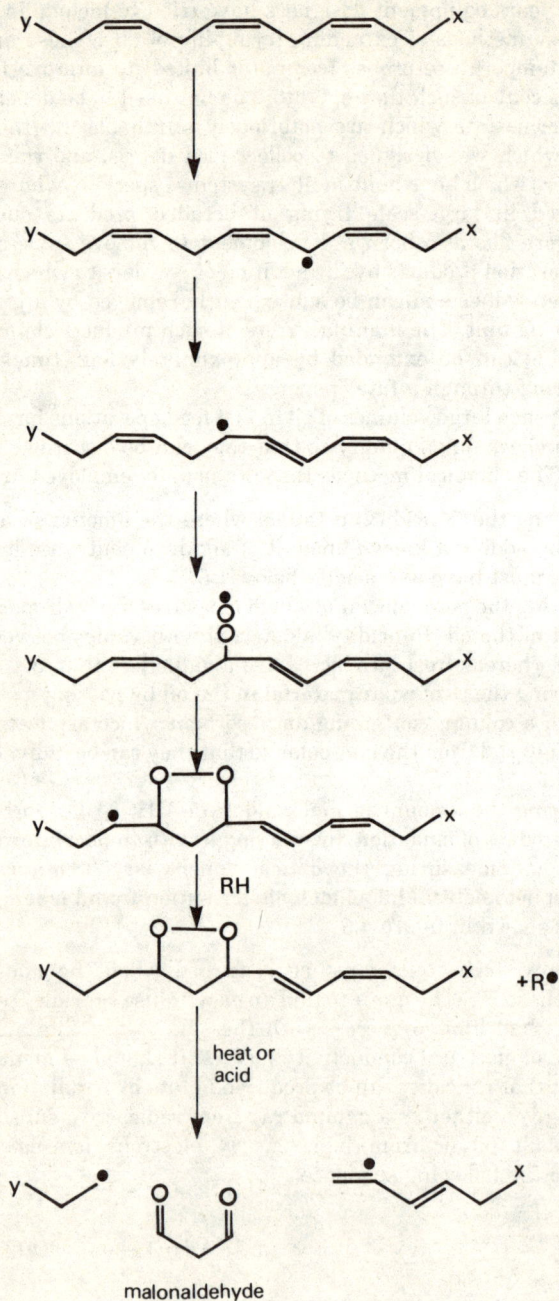

4.5   *The formation of malonoldehyde from linolenic acid*

Fats and oils do not have a fixed melting point as does water because they are mixtures not sole substances. The range of temperatures over which the fat melts depends on the fatty acid composition of the triglycerides. It can be seen from table 4.2 that the lower melting points are obtained with shorter fatty acid chains and increasing degree of unsaturation.

Table 4.2 *Melting points of fatty acids*

Fatty acid	Symbol	Melting point °C
Butyric	$C_{4:0}$	−8
Lauric	$C_{12:0}$	44
Palmitic	$C_{16:0}$	63
Stearic	$C_{18:0}$	70
Oleic	$C_{18:1}$	11
Linoleic	$C_{18:2}$	−5
Linolenic	$C_{18:3}$	−11

This ability to change over a range of temperatures from a liquid to a solid gives a fat plasticity. This plasticity is important when fats are used at room temperature for spreading, creaming and in pastry manufacture. It is possible to buy various margarines, which are products made from the hydrogenation of vegetable oils, to suit the particular requirements of the products in which they are to be used. Puff pastry contains a slightly higher melting point margarine, extending over a wider range, than short pastry margarine, in order to prevent the layers of fat melting during its manufacture. The amount of unsaturation in a fat is measured by the iodine value. Iodine is taken up by fats and oils such that the more unsaturated the fat the higher the iodine value. The value is expressed in grams of iodine absorbed by 100 g of fat. Typical values are 80–90 (olive oil), 47–67 (lard) and 35–45 (beef dripping). This test is often carried out if a pure fat is thought to have been mixed with another and therefore not meeting the requirements of the description required by law.

Oils and fats also form important emulsions in catering such as mayonnaise and hollandaise sauces and their derivities. The manufacture of such emulsions requires the dispersion of the fat into a water environment and the subsequent avoidance of separation as occurs with unstable emulsions such as vinaigrette. Fats and oils do not disperse naturally in water because they are chemically opposite to each other. This chemical tension results in the two phases making minimal contact with each other when placed together in a container. An input of energy by vigorous whisking can overcome this and so the oil can be dispersed in the water as droplets. If the energy is removed the droplets coalesce and the oil and water separate into their respective phases. However, if an emulsifying agent is added at the beginning of the whisking stage the emulsion can be stabilized. This stabilizing property of emulsifying agents or surface active agents resides in the fact that their molecules have two regions, one liking water (*hydrophilic*) and one liking fat (*lipophilic*). The effect is to surround the oil drops with an emulsifying agent which prevents coalescence and to reduce the surface tension. The common emulsifying agents are *lecithin* – a phospholipid found in egg yolk, *glycerol monostearate* – an artificial emulsifer used in salad dressings, *proteins*, such as casein, found in milk and, although not food molecules, soaps and detergents are surface active agents. Emulsifying action and emulsifying agents are shown in figure 4.6.

The amount of fat in food has become one of the major nutritional aspects of 'healthy eating'. It is therefore important from this point of view, and from the fact that products may be subjected to fat labelling, to be able to measure the % fat in foods. There are several ways of doing this:

(1) By using the Soxhlet apparatus where fat is washed out of the food using petroleum spirit as solvent.
(2) If the investigation is to measure fat absorption during frying then the % food solids after evaporation, before and after frying, can be compared to find the % fat absorbed.
(3) If a sample is digested in a sulphuric acid amyl alcohol mixture, the liberated fat that separates out can be measured (Gerber method).

CARBOHYDRATES

The carbohydrates can be divided into the monosaccharides, disaccharides and the polysaccharides. The monosaccharides and disaccharides form the sugars and the polysaccharide group includes the starches, pectin and cellulose. The basic building brick of the carbohydrates is the monosaccharide.

Monosaccharides such as glucose, fructose and galactose all have the chemical formula $C_6H_{12}O_6$. They differ from each other in their spacial arrangement (*isomers*) and when combined together in pairs they form the disaccharides and, in their hundreds, the polysaccharides. All the monosaccharides carry out chemical reduction with Fehling's solution where the valency of copper is reduced from two to one indicating that the possible monosaccharide functional group to be an aldehyde. However, all aldehydes give a pink colour with Schiff's reagent and the monosaccharides do not, so the original structure of an open chain was changed to one of the cyclic compound as shown in figure 4.7. There are two forms of glucose α and β and they differ in having a different orientation at carbon 1 with the α form having its OH group below the plane of the ring. The structure of galactose is very similar to α-glucose except the OH group is up at carbon 4 whereas fructose can exist as a six-membered ring or, as shown in figure 4.7, as a five-membered ring.

The *disaccharides* sucrose, maltose and lactose vary because they are composed of different monosaccharides; maltose is two glucose

an emulsion

lecithin

glycerol monostearate

alkyl benzene sulphonate

quaternary ammonium compound

soap

non-ionic detergent

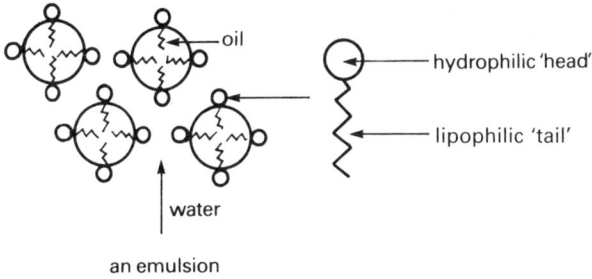

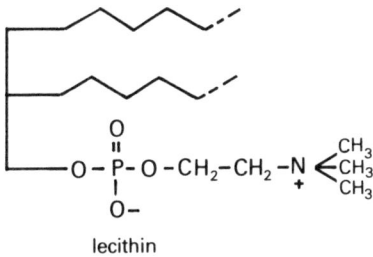

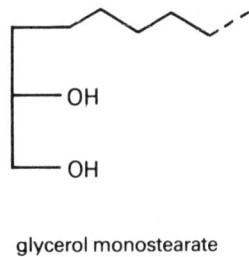

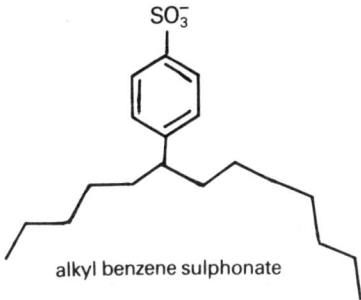

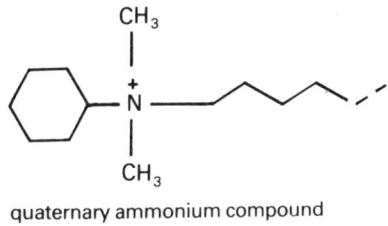

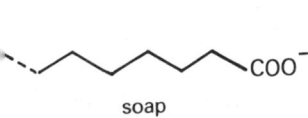

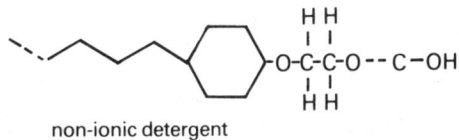

4.6 *Emulsifying or surface active agents*

molecules, sucrose is a glucose and a fructose molecule whereas lactose
is glucose and a galactose molecule. The two monosaccharides are
joined together in the disaccharides by the removal of a molecule of
water from two of the OH groups. The linkage is different for each of
the disaccharides, as can be seen in figure 4.7, but can also be
demonstrated by the fact that their hydrolysis requires three different
enzymes; sucrase, maltase and lactase. The power of chemical

reduction resides in carbon 1 for glucose and galactose and in carbon 2 for fructose. The disaccharide sucrose is therefore not capable of reducing Fehling's solution because the 1–2 linkage involves both reducing positions. When testing for sucrose in foods the sugar needs to be hydrolysed first using an acid as catalyst to its two monosaccharides. This mixture of glucose and fructose is known as *invert sugar* because of its behaviour towards polarised light.

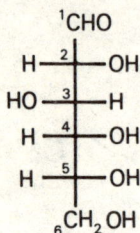

glucose

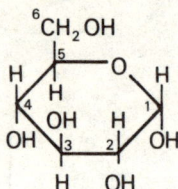

∝ glucose

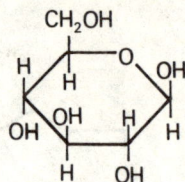

ß glucose

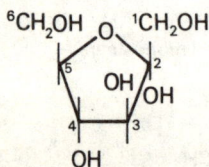

fructose

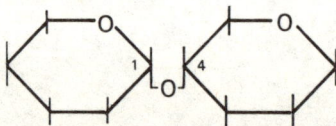

maltose

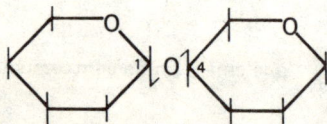

lactose

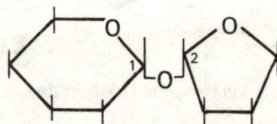

sucrose

4.7   The sugars

Glucose is often referred to as dextrose or grape sugar. It is a white solid that does not crystallize easily and is obtained by the hydrolysis of starch. It is not as sweet as sucrose (0.7 where sucrose = 1). Fructose is often referred to as fruit sugar and is found with glucose in the juice of sweet fruits and honey. It is a ketone in its chain form and is obtained when sucrose is hydrolysed. Like glucose it does not crystallize easily and is much sweeter than all the sugars (sweetness = 1.7).

Sucrose is the major sugar and because of its easy accumulation by bacteria on the surface of teeth has been associated with dental caries as well as obesity. The products developed from sucrose are basically crystalline and non-crystalline (amorphous). In sugar cookery fondant sugar is a crystalline product with a smooth texture due to the minute crystal size. The development of crystals requires the formation of nuclei and the deposition of molecules on these nuclei. Crystal growth after the formation of stable nuclei involves several steps: (1) transport from the medium to the growing crystal; (2) adsorption on the crystal surface; and (3) orientation in the surface. Sucrose requires a considerable degree of super saturation before crystallization commences as nuclei form more readily in this concentrated solution. However, the high viscosity of such solutions hinders the transport of the sucrose molecules to the crystal surface and therefore the higher the boiling point of the syrup (145°C) the less likely the formation of crystals. The most favourable temperature for crystal growth of a saturated sucrose solution boiled to 115°C is between 70°C and 90°C and so to avoid large crystal formation the syrup is allowed to cool before beating which favours nuclei formation but hinders deposition. Impurities such as glucose and fructose become adsorbed onto the crystal face but cannot be orientated into the crystal structure thus slowing down or stopping crystal growth. Fondant icing is thus prepared with sucrose, glucose, heating to 115°C and beating when cool whereas hard sugar glazes are obtained by sucrose, acid, boiling to 145°C and allowed to cool without beating. If sugar is boiling to 175°C the molecules decompose to produce a brown bitter material known as *caramel*.

The disaccharides maltose (sweetness = 0.3) is obtained by enzyme action on starch. Lactose (sweetness = 0.15) is only found in milk.

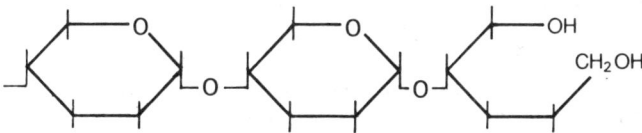

hydrogenated glucose syrup

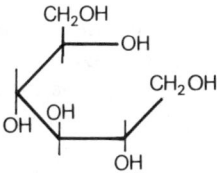

sorbitol

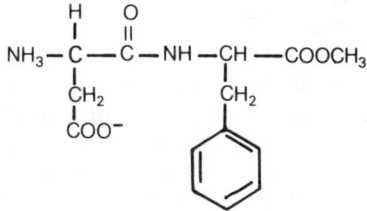

aspartame

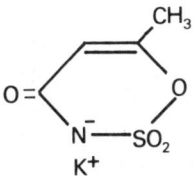

acesulfame K

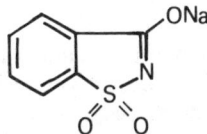

sodium saccharin

4.8 Sweeteners

Nutritive sweetners to replace sucrose have been in existance for over 100 years and their structures are shown in figure 4.8. Saccharin was first synthesized in 1879 by Fahlberg who by accident found it had a sweet taste. Sodium saccharin is essentially non-caloric and has a very sweet taste (sweetness = 450). It is a stable compound with respect to heat and time, can be synthesized very cheaply with few impurities, has a synergism with respect to sweetness when combined with other sweeteners but has its problems with its well-known after taste and concern regarding its safety. An article on saccharin and its problems is outlined in the *Journal of The American Dietetic Association*, July 1986, Volume 86 No. 7, pages 929–931. In response to the banning of cyclamate and the concern over saccharin several new sweeteners have appeared recently on the market. The replacement for sugar to reduce dental decay has resulted in the production of sorbitol and hydrogenated glucose syrups. Sorbitol (sweetness = 0.7) is less sweet than sucrose and is being used to replace sucrose in wort to produce low alcohol beverages, and in a range of confectionery. Inulin, which is a fructose containing polysaccharide, found in Jerusalem artichokes, is also under consideration to provide fructose syrups. These sweeteners are metabolised by the body and so give the same calorific value as any carbohydrate (approximately 17 J/g). Aspartame is probably the most successful of all the new sweeteners and goes under the brand name of *Nutrasweet*. Aspartame (sweetness = 200) is sweeter than sucrose and is a dipeptide of two amino acids phenylalanine as the methyl ester and aspartic acid. It is, therefore, a natural food and although it has an energy value of 17 J/g it is used in such small amounts to have only minute effects on the final product energy. Aspartame is safe, with a clean sweet taste without any metallic after taste and can be used in liquids and table top sweeteners. On digestion 19 mg of aspartame (equivalent to a spoonful of sucrose) is metabolised to 7.6 mg of aspartic acid, 9.5 mg of phenylalanine and 1.9 mg of methanol. This is to be compared with a glass of milk which produces 528 mg of aspartic acid and 542 mg of phenylalanine, and a glass of tomato juice which produces 47 mg of methanol. The only nutritional drawback is that its contents need to be made aware of to people who suffer from the disease phenylketouria. In food processing it is used without restrictions in a great number of foods except those requiring prolonged high temperature such as baking or frying. Due to its amide and ester bonds which are subject to hydrolysis at temperatures above 150°C it suffers from a loss of sweetness and finds its greatest stability between pH 3 and 5.

Acesulfame K (sweetness = 200) is a very soluble material which is much sweeter than sucrose. The onset of sweetness is fast and persists longer than sucrose. It is fairly stable down to a pH 3.0 and acceptable down to pH 2.5, and in baking trials has shown temperature stability up to 220°C. It is not metabolised and appears to be fairly easily removed from the body. These sweeteners and others in development are going to enable the reduction of sucrose to take place in our diet and with the possibility of using them together reduces the possible harm that any one of them could have.

The **polysaccharides**, as their name suggests, are large molecules with the monosaccharides as their building bricks. Starches such as wheat, arrowroot, maize, rice and potato are all polysaccharides where the building brick is ∝ glucose. The starches differ in having a different composition of the two basic starch molecules, the straight chain amylose and the much branched amylopectin. Table 4.3 shows the composition of some natural starches.

Table 4.3 *The composition of some natural starches*

Starch	% Amylose	% Amylopectin
Maize	24	76
Wheat	22	78
Potato	20	80
Arrowroot	20	80
Rice	16	84

In the older classification high amylose starches such as maize were known as the *non-waxy* starches whereas the low amylose starches such as rice were known as the *waxy* starches. When starches are dispersed in water and heated the viscosity of the liquid increases as the temperature approaches boiling point, falls and then rises again. These changes can be seen in figure 4.9 for potato and rice starch where (a) the rise in viscosity is due to the swelling of the granules; (b) the bursting of the granules with the release of the starch molecules decreases the viscosity; and (c) the formation of the starch gel increases the viscosity as the temperature falls.

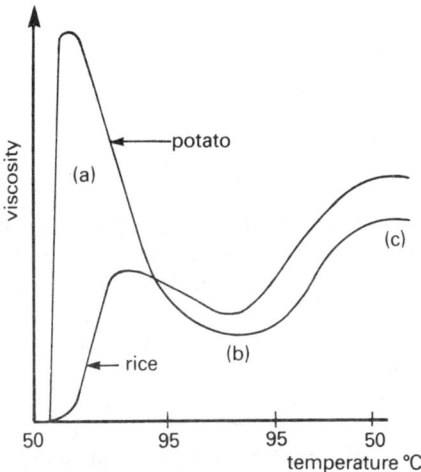

4.9 Viscosity of starch during cooking

50   *Food Molecules*

Gels are dispersions in water that are not solutions or suspensions but colloids. Colloids differ from suspensions and solutions according to the properties outlined in table 4.4.

Table 4.4 *Solutions, colloids and suspensions*

Solutions	Colloids	Suspensions
Small molecules or ions	Macromolecules (polymers)	Too large to remain dispersed
Particle size $< 10^{-8}$m	Particle size $10^{-8} \longrightarrow 10^{-6}$m	Particle size $> 10^{-6}$m
No gels	Gels	No gels
Transparent	Translucent	Opaque
Intense motion	Little motion	No motion
High osmotic pressure	Low osmotic pressure	No osmotic pressure

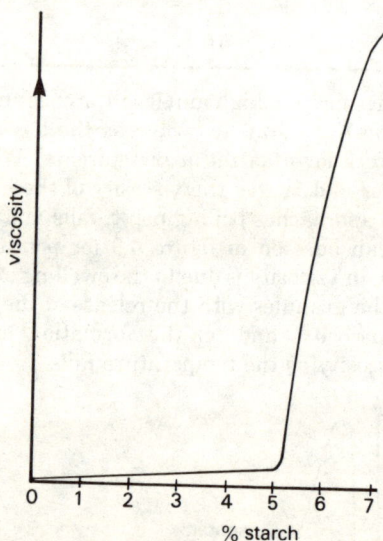

4.10   Viscosity of potato starch gels at 20°C

Starch molecules released from the granules on heating are large molecules and, according to the kinetic theory, stationary compared with water molecules. The formation of a gel can be looked upon as the reduction in overall mobility of water molecules by some of them becoming attached to the larger starch molecules by hydrogen bonding. The viscosity therefore increases with the amount of starch added and by a reduction in temperature. This increase in viscosity with increase in the amount of starch added can be seen in figure 4.10. However, it can be seen that the viscosity increases very sharply between a concentration of 5% and 6% which cannot be explained simply on immobilisation of water molecules. These firmer gels are thought to be the result of starch molecules being in sufficient

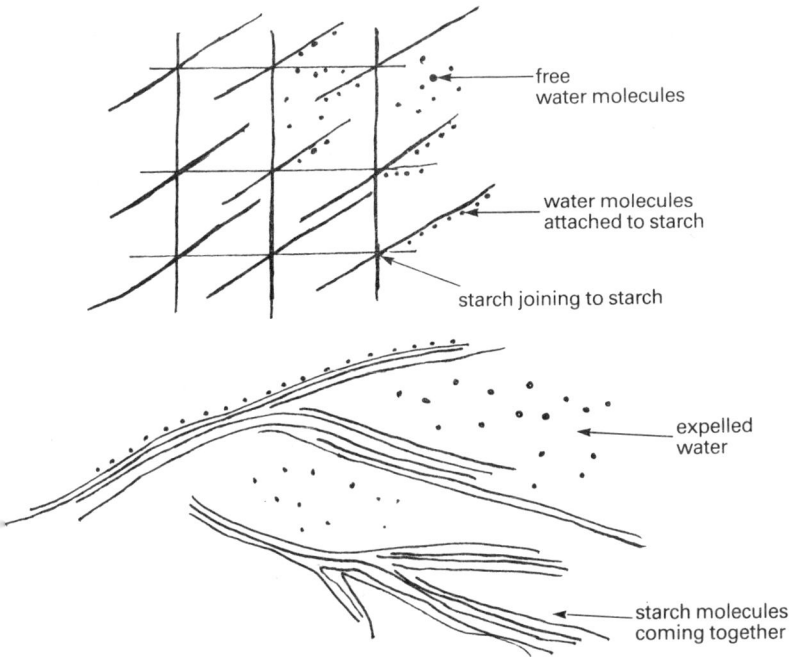

4.11   Starch gels before and after retrogradation

numbers to be able to form a scaffolding or lattice by joining together in parts as shown in figure 4.11. The ability of a starch to form a firm gel, such as blancmange, will depend on the ease of formation of this lattice which is better achieved with straight chains than with branched. Maize starch firm gels are formed at a lower concentration than rice firm gels and some highly branched modified starches will not form a firm gel at all. The stability of a starch gel depends on a number of factors:

1   *Time* – within a few days firm starch gels such as trifle custards begin to weep (*syneresis*). This loss of water is due to starch molecules coming together and forcing the water off (*retrogradation*).

2   *Acidity* – when starch gels are used to thicken fruit pie fillings or lemon meringue pie the acid causes hydrolysis of the starch to sugar reducing the strength of the gel.

3   *Sugar* – the addition of sugar causes a competition for water between the starch and itself. Firm starch gels will not form in high sugar concentrations.

4   *Egg yolk* – the addition of egg yolk to soups, cakes, choux paste and duchess potatoes can cause hydrolysis of the starch due to the presence of the starch digesting enzyme amylase. This enzyme can turn a velouté soup to a thin liquid unless the enzyme (resistant up

to 70°C) is denatured by returning the soup to the boil (cook out) after the addition of the liaison.

5 *Freezing* – the freezing of starch gels during cook-freeze catering operations results in retrogradation of starch gels. This usually results in excessive weeping from the firm gels and small particles of starch in the thin gels of soups and sauces. To reduce this, highly branched modified starches (freeze and flow) which do not easily come together are used to replace a part of the natural starch in the recipe.

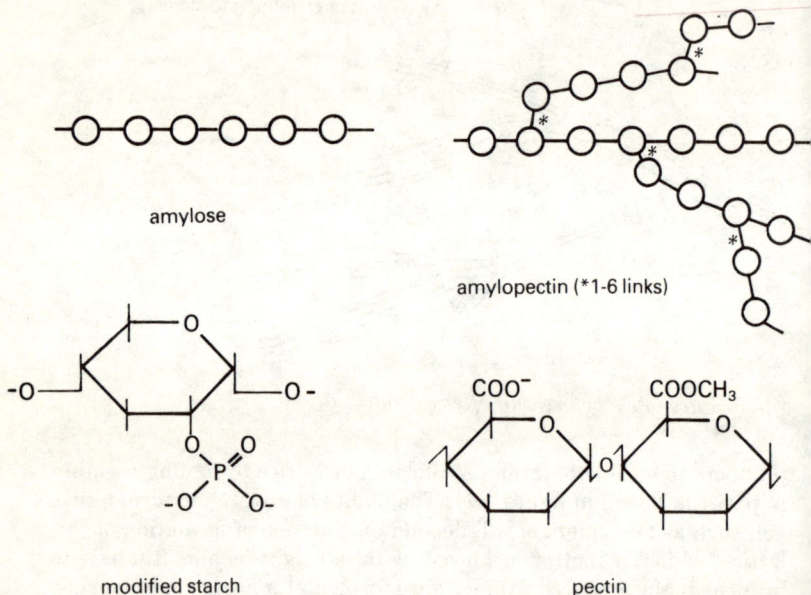

amylose

amylopectin (*1-6 links)

modified starch

pectin

4.12 *Polysaccharides*

Pectin is a polysaccharide formed from two building bricks based upon the monosaccharide galactose, galacturonic acid and its methyl ester as shown in figure 4.12. The difference between pectin and starch is that pectin is found in the cell wall of plants and because of its composition can have a varying negative charge and therefore differences in gel formation. In some cases, for the lattice to form, the negative charge needs to be removed as like charges repel and, as pectin has such an affinity for water, the lattice can be prevented from forming unless some of the water is removed by the presence of sugar. The properties of pectin gels are shown in table 4.5.

Starches and pectins give a blue black colour with iodine indicating that the polymer is not straight but forms a helical structure. This open structure accounts for the ability of starch and pectin to absorb such large amounts of water. Cellulose is a polysaccharide made from the monosaccharide β-glucose. It is not digested by the enzymes of the digestive tract in man like pectin and forms a part of dietary fibre. It

Table 4.5 *Pectin gels*

Type of pectin	Gelling properties
Completely methoxylated (neutral charge)	Requires sugar but no acid
70% to 80% methoxylated	Requires sugar and acid pH 3 to 3.4
50% to 70% methoxylated	Requires sugar and acid pH 2.8 to 3.2
Low methoxylated pectin	Gels formed with cations (Ca^{++}) for diabetic jams

does not give a blue black colour with iodine which accounts for its ability not to form gels but still has the property of attracting some water which gives the comfort associated with cotton clothes. The other constituents of dietary fibre besides cellulose and pectin are the other undigested parts of the plants cell wall: the hemicelluloses and lignin. The *hemicelluloses* contain any number of kinds of hexose and pentose monosaccharides whereas *lignin*, which is not a carbohydrate, has a chemical similarity to the flavanoid compounds which form the water-soluble red, blue or purple pigments.

PROTEINS

Proteins are polymers and, like starches (gels) and fats (emulsions), are able to form colloidal dispersions (gels and foams). The building bricks for proteins are 20 or so amino acids with the basic structure outlined in figure 4.13 and the different R groups in table 4.6.

Table 4.6 *The R groups for some amino acids*

Amino acid	R	Amino acid	R
Glycine	H	Serine	CH_2OH
Alanine	CH_3	Threonine	$CH\,CH_3OH$
Valine	$CH(CH_3)_2$	Aspartic acid	CH_2COO^-
Leucine	$CH_2CH(CH_3)_2$	Glutamic acid	$CH_2CH_2COO^-$
Phenylalanine	CH_2-⬡	Lysine	$(CH_2)_4NH_3^+$
Cysteine	CH_2SH	Tyrosine	CH_2-⬡ OH

The carboxyl group being acidic can lose a hydrogen ion and become negative whereas an amino group can pick up a hydrogen ion and become positive. These differently charged R groups account for the fact that proteins have different charges distributed on their surfaces. The protein is made by joining together amino acids with the loss of water to form the peptide bond (figure 4.13).

Proteins have many functions to perform within the cells of living plants and animals. These functions involve providing pigments for

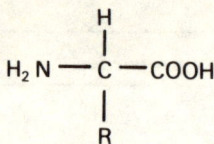

amino acid

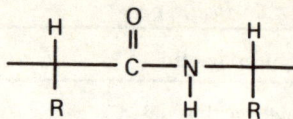

peptide bond

4.13   Amino acids and the peptide bond

the transport and storage of oxygen (haemoglobin and myoglobin), the contractile mechanism in muscle (actin and myosin) and the enzymes which carry out the metabolic processess. In order for all these functions to be carried out the proteins take up different three-dimensional shapes such that each protein is different from another. This three-dimensional structure needs to be reinforced by extra bonds – the hydrogen bond and the disulphide bond. The *hydrogen bond* is a weak bond easily broken by an increase in temperature or mechanical/electrical forces, whereas the *disulphide bond* is a covalent bond only broken by chemical reduction. The bonding can be seen in figure 4.14 where the hydrogen bond forms from the remainder of the carboxyl and amino group and the disulphide bond between the R groups of the amino acid cysteine. The hydrogen bonding is the most abundant whereas the disulphide bond is small in number because it relies upon the close proximity of two cysteine molecules.

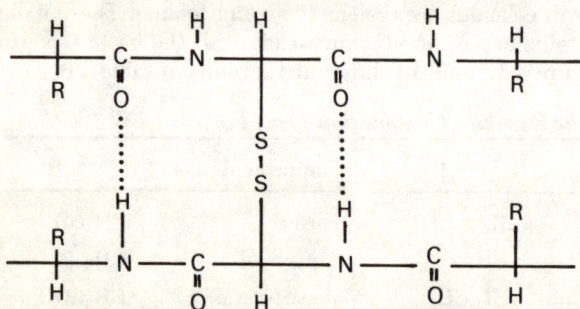

4.14   Hydrogen and disulphide bonds in proteins

If the protein molecule takes up a globular form, as shown in figure 4.15, it displays regular internal structures such as helices and as it is usually dispersed in water has its polar R groups on its surface. The overall charge on the surface of a protein is negative and this property allows mixtures of proteins to be separated by attracting them to a positive electrode. Heating, beating or changing the ionic nature of the environment causes proteins to uncoil or denature. This loss of shape results in the proteins ability to form a gel or a foam and in the case of an enzyme to lose its activity.

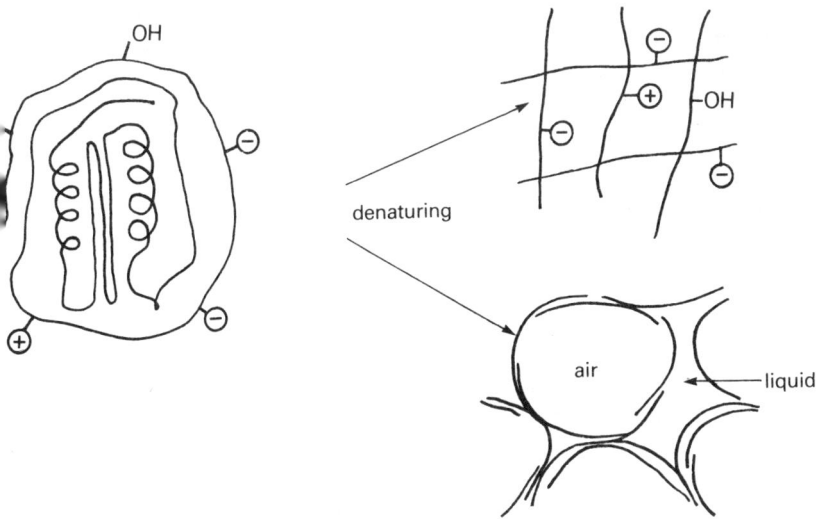

4.15   Globular proteins forming gels and foams

When proteins are uncoiled the non-polar R groups are exposed to the watery environment and the tendancy for proteins to aggregate together again is much stronger than in polysaccharides. The formation of a gel in for instance an egg custard or milk junket is probably due to the fact that the overall negative charge on the protein prevents this retrogradation by electrostatic repulsion. This overall negative charge can be removed by the addition of positive ions such as H^+ ions, and the pH at which the overall charge on a protein is zero is known as the *iso electric point*. For egg white proteins it is approximately pH = 4.8 and for milk proteins pH = 5.2. The result on milk and egg proteins can be seen by the addition of acid which results in milk proteins clotting to form yoghurt and cheese and a tighter white which is obtained when poaching eggs in water containing vinegar.

The formation of an egg white foam relies upon proteins coming together to form the skin of the air sacs. This negative charge is therefore often removed by the addition of the acid, cream of tartar. It is also easy to see why fat from egg yolk causes loss in foaming ability because the non-polar triglycerides are attracted to the non-polar R groups preventing protein alignment. This ability of a protein to exhibit non-polar and polar properties on the same molecule has given proteins an emulsifying role in natural products such as milk.

Milk proteins consist mainly of α-casein which is already denatured for ease of digestion but fails to form a gel because it is contained in packets surrounded by another protein, k-casein. This outer coating can be removed by digestion using rennet or the use of an acid to release the α-casein to form a gel (junket) or a clot (cheese) depending on the pH. Meat proteins are not present to provide food but to act as an oxygen store and provide the contractile system. Myoglobin provides the muscle with a supply of oxygen and gives it the red colour. It is a conjugate protein in that it has more than an amino acid chain. The addition to the normal protein structure is a haem group which consist of four pyrrole rings and an iron atom as shown in figure 4.16. The various attachments to the spare arm on the iron atom and its change in valency account for the various colours of meat. The other meat proteins involve collagen and elastin of the connective tissue, and actin and myosin of the contractile system. It can be seen from figure 4.16 that these proteins are linear because they are forming fibres.

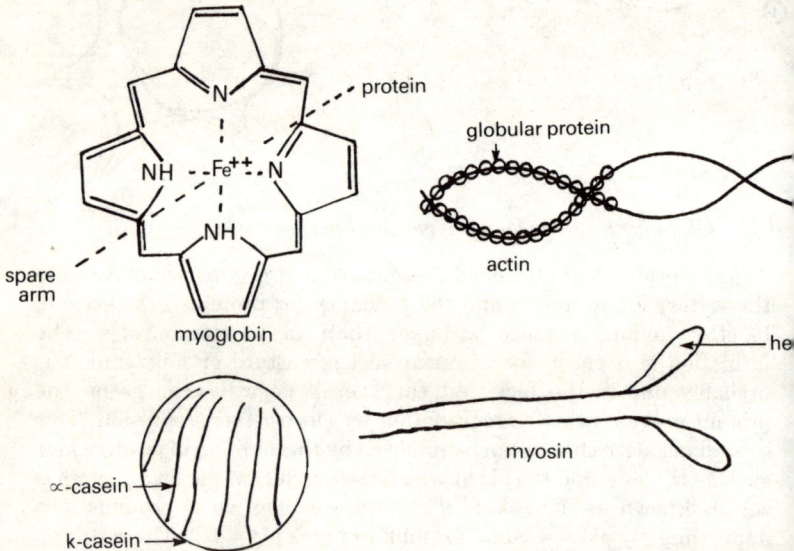

4.16   Meat and milk proteins

When foods are cooked, amino acids and proteins react with reducing sugars to form a variety of compounds which give foods flavour, odour and a dark colour. These reactions are known as *Maillard reactions* and occur in most cooking situations except boiling, steaming and microwave. The reactions do not occur to any great extent at temperatures below 100°C, they prefer an alkaline environment and require a reducing sugar not sucrose. In the baking of bread the development of the colour of the crust can be inhibited if the reducing sugar content from the flour is low. The development of flavour and colour of meats by frying before they are used in casseroles and the roasting of coffee beans are all positive developments of Maillard reactions.

NUTRITION

In order to survive we need to provide ourselves with sufficient nutrients in our diet. The nutrients can be classified into carbohydrates and fats which provide our main energy requirements, proteins which provide replacement materials such as new skin and the less abundant nutrients the vitamins and minerals.

The recommended daily amounts (RDAs) of some of the population groups are shown in table 4.7. In menu planning it is not necessary to provide the precise requirements each day as the nutrients can be stored to a varying extent. An analysis covering a few days is adequate. In analysing meals for nutritional requirements it is necessary to consult food tables. The information given in table 4.8 is extracted from *McCance and Widdowson: Composition of Foods* by Paul and Southgate (HMSO) and is going to be used to calculate the amount of energy in kilojoules in the following two meals:

MEAL 1   Chicken chasseur (1 portion)
1 portion of chicken (3 lb chicken for 4 people)
Mushrooms   25 g
Dripping   10 g
Flour   12 g
Onion   12 g
Tomatoes   50 g
Wine   15 g
Butter   10 g
⅓ can new potatoes (can of new potatoes = 450 g)
1 portion of spinach with 10 g of butter (1 lb of fresh
spinach for 3 portions)
1 tangerine   60 g
1 cup instant coffee (black)   5 g

MEAL 2   Chicken meal on the plate (1 portion)
Sliced roast chicken   100 g
Canned potatoes   100 g
Boiled spinach   80 g
Butter   10 g
Fresh fruit salad of:
    Tangerines   50 g
    Raspberries   50 g
    Strawberries   50 g
1 cup of coffee infusion (black)   150 ml
The analysis for the two meals is set out in table 4.9.

The following notes should accompany the analysis:
1   The 3 lb chicken (1362 g) is divided into four portions = 340 g.
2   The edible portion (do not eat the bone) is 0.64.

3   The chicken is eaten with the skin, so the energy value per g = 9.54 kJ.
4   The tables do not show edible waste when tomatoes have their skin, seeds and juice removed. The energy value should be less but it is still only small.
5   Canned potatoes contain a liquid that is drained off.
6   MEAL 1 needs to have wastage taken into account because the preparation begins with fresh produce. MEAL 2 is as the meal appears on the plate and wastage does not need to be considered unless of course it is not all eaten.
7   You can see it is tedious to do the calculations by hand and so you can purchase computer programmes such as 'Superdiet' from Super Software Systems Ltd to do the work for you.

The protein intake must supply sufficient quantities of the essential amino acids. These amino acids cannot be obtained by the metabolism of any of the other so called 'non-essential amino acids' to any great extent. They must therefore be in the dietary proteins and if a protein

Table 4.7   *RDAs of food energy and some nutrients for some population groups in the UK (DHSS 1979)*

Age ranges (years)	Energy MJ	Carbohydrate[1] g	Fat[2] g	Protein g
Boys				
3–4	6.5	210	61	39
7–8	8.25	265	77	49
12–14	11.0	350	103	66
15–17	12.0	385	112	72
Girls				
3–4	6.25	200	58	37
7–8	8.0	255	74	47
12–14	9.0	285	84	53
15–17	9.0	285	84	53
Men				
18–34 Moderately active	12.0	385	113	72
Very active	14.0	445	130	84
35–64 Moderately active	11.5	365	107	69
Very active	14.0	445	130	84
65–74 Sedentary	10.0	320	93	60
Women				
18–54 Most occupations	9.0	285	84	54
Pregnancy	10.0	320	93	60

(1)   Weight required to provide at least 50% of the energy.
(2)   Weight required to provide no more than 35% of the energy.

were to contain all the amino acids in the right proportions it would have a biological value of 1.0. The biological value of a protein is:

$$\text{Biological value} = \frac{\text{Amount of protein retained in the body}}{\text{Amount of protein absorbed from the diet}}$$

The biological values of eggs and human milk are the highest (0.9–1.0), meat, fish and cows' milk (0.75–0.8), soya (0.7), wheat (0.5) and peanuts (0.4–0.45).

The essential amino acids for an adult are leucine, isoleucine, lysine, methionine, valine, phenylalanine, threonine and tryptophan with histidine for the growth of infants. The western diet normally involves the consumption of a wide range of protein foods and so the essential amino acids are provided in full. This nutritional feature is known as *complementation* but it is advisable to note that if the range of proteins is restricted, for example in vegetarian and vegan diets, then there is a possibility of a lack of some if not all essential amino acids, eg cereal proteins are low in lysine and soya is low in methionine.

The nutritional lossess from food during cooking is often quite small for the main nutrients such as carbohydrates, fats and proteins, and mineral losses are not often considered to be of major significance.

Continued

Thiamin mg	Ascorbic acid mg	Vit A µg (retinol equivalents)	Calcium mg	Iron[3] mg
0.6	20	300	600	8
0.8	20	400	600	10
1.1	25	725	700	12
1.2	30	750	600	12
0.6	20	300	600	8
0.8	20	400	600	10
0.9	25	725	700	12
0.9	30	750	600	12
1.2	30	750	500	10
1.3	30	750	500	10
1.1	30	750	500	10
1.3	30	750	500	10
1.0	30	750	500	10
0.9	30	750	500	12
1.0	60	750	1200	13

(3) Iron recommendations do not cover heavy menstrual losses.

Table 4.8   *The composition of some foods per 100 g*

Food	kJ	Protein	Fat	Starch	Sugars	Edible matter proportion of weight purchased
Chicken raw						
meat only	508	20.5	4.3	0	0	0.44
meat and skin	954	17.6	17.7	0	0	0.64
Chicken roast						
meat only	621	24.8	5.4	0	0	0.40
meat and skin	902	22.6	14.0	0	0	0.55
Dripping	3663	0	99.0	0	0	1.0
Butter	3041	0.4	82.0	0	0	1.0
Flour white (plain)	1493	9.8	1.2	78.4	1.7	1.0
Mushrooms (raw)	53	1.8	0.6	0	0	0.75
Onions (raw)	99	0.9	0	0	5.2	0.97
Tomatoes (raw, skin, seeds)	60	0.9	0	0	2.8	1.0
Spinach (leaves, boiled)	128	5.1	0.5	0.2	1.2	0.42
Potatoes (canned)	226	1.2	0.1	12.2	0.4	0.63
Tangerine (raw)	143	0.9	0	0	8.0	0.70
Raspberries (whole raw)	105	0.9	0	0	5.6	1.0
Strawberries (raw, no stalks)	109	0.6	0	0	6.2	0.97
Coffee (infusion)	8	0.2	0	0.3	0	1.0
Coffee (instant)	424	14.6	0	4.5	6.5	1.0
Wine (medium white)	311	0.1	0	0	3.4	1.0

However, substantial losses in vitamin content can occur that is not reflected when using the nutritional tables. Water soluble and heat labile vitamins such as vitamin C (ascorbic acid) can in certain circumstances be completely lost from the food especially in conditions of hot holding as can be seen in table 4.10.

In 1984 the COMA report (Committee on Medical Aspects of Food Policy) issued by the DHSS reported on *Diet in Relationship to Cardiovascular Disease*. It had been ten years since the previous report on *Diet and Coronary Heart Disease* and after reviewing considerable scientific data a balance of evidence was found to justify a recommendation for dietary change. The recommendations were mainly concerned with dietary changes to decrease the incidence of coronary heart disease (CHD) and the reduction of high blood pressure (main risk factor for cerebrovascular disease) which is increased by obesity, high alcohol intake and may be by the intake of common salt.

CHD and cerebrovascular disease cause 40% of death in men and 38% of death in women. The UK has not yet seen the declines in mortality from CHD enjoyed by the USA, Canada, Australia, New Zealand, Belgium and Finland.

Table 4.9 *Analysis of meals*

Food	Weight g	Edible weight g	Energy kJ
MEAL 1			
Chicken	340	340 × 0.64 = 218	218 × 9.54 = 2079
Mushrooms	25	25 × 0.75 = 19	19 × 0.53 = 10
Dripping	10	10	10 × 36.63 = 366
Flour	12	12	12 × 14.93 = 179
Onion	12	12 × 0.97 = 12	12 × 0.99 = 12
Tomatoes	50	50	50 × 0.6 = 30
Wine	15	15	15 × 3.11 = 47
Butter	10	10	10 × 30.41 ≐ 304
Canned potatoes	150	150 × 0.63 = 95	95 × 2.26 = 214
Spinach	151	151 × 0.42 = 64	64 × 1.28 = 81
Butter	10	10	10 × 30.41 = 304
Tangerine	60	60 × 0.7 = 42	42 × 1.43 = 60
Coffee	5	5	5 × 4.24 = 21
		Total	3707 kJ
MEAL 2			
Roast chicken (no skin)	100	100	621
Potatoes	100	100	226
Butter	10	10	304
Spinach	80	80	102
Tangerine	50	50	72
Raspberries	50	50	53
Strawberries	50	50	55
Coffee	150	150	12
		Total	1445 kJ

In world ranking of death caused from CHD in 1978 the countries of the UK occupied three of the top five positions for men and two of the top five positions for women.

Diet in the UK from 1952 to 1982 has shown:

1   A decline in the total energy consumed over the period stated.
2   Protein in the diet has remained fairly constant (11%).
3   The % energy from fat has risen from 37% to 42% and is not decreasing.
4   Carbohydrate intakes have shown the opposite trend of decreasing from about 51% of total energy to about 45% in recent years.
5   The ratio of polyunsaturated to saturated fatty acids (P/S ratio). The P/S ratio has increased from 0.17 to 0.27 from 1959 to 1982.
6   The dietary fibre has remained more or less constant at around 20 g per day.
7   There was no information on the amount of salt consumption.

Table 4.10   *Vitamin C lost from frozen peas during cooking and hot holding*

Time (mins)	Procedure	Ascorbic acid mg/100 g
0	Frozen peas put in boiling water	20.5
35	Water began to boil	15.0
50	Peas strained and put on hot plate	8.1
85	Peas taken from hot plate	3.7
145	Peas served to customers	1.1

The recommendations are based upon the strong positive relationship between the amount of dietary saturated fatty acids and death from CHD. A positive relationship has also been found between total blood cholesterol and CHD and that saturated fatty acids increase blood cholesterol whereas dietary polyunsaturated fatty acids lower blood cholesterol. The recommendations are that the consumption of saturated fatty acids and fat in the UK should be decreased. There is also a recommendation that the ratio of polyunsaturated to saturated fatty acids (including the trans polyunsaturated which have much higher melting points than the cis polyunsaturated fatty acids) P/S ratio be increased to approximately 0.45. The intakes recommended are 15% of food energy for saturated fats plus trans polyunsaturated and 35% of total energy for total fat. The recommendations were not intended for children below five or for people who already have an appropriate diet. The report gave no specific recommendations concerning the dietary intake of cholesterol, sugars, alcohol, salt or dietary fibre.

The NACNE discussion paper on proposals for Nutritional guideline for health education in Britain was published in 1983 and gave further recommendations. The report highlighted the prevalence of excess weight which is shown by a vary substantial portion of the population. The acceptable weights can be calculated from the Body Mass Index (BMI) which for men should be between 20.1 and 25 (obese = 30) and for women between 18.7 and 23.8 (obese = 28.6).

$$\text{BMI} = \frac{\text{Weight in kg}}{\text{Height}^2 \text{ in metres}}$$

The long term proposals are that fat intakes should be on average 30% of the total energy intake. Saturated fatty acids should be on average 10% of the total energy intake but no recommendations about the P/S ratio as the other recommendations will automatically increase it. The sucrose intake should be reduced to 20 kg per head per year from the present 38 kg per head per year (a reduction of approximately 50%). A reduction of salt and alcohol by 10% and an increase in dietary fibre from cereals, fruits and vegetables to 30 g per day.

The changes outlined above are not vast alterations but fairly small changes easily achievable by individuals or those in the food production industries. The COMA report gives information on how to achieve the recommendations for dietary change. Table 4.11 shows the major sources of saturated and polyunsaturated fatty acids and of total fat in the diet in Great Britain in 1981.

Table 4.11 *Major sources of fats and fatty acids in the GB diet (1981)*

			Fatty acids			
	Fat		Saturated		Polyunsaturated	
Food	g/day	% total	g/day	% total	g/day	% total
All milk, cream	14.5	14.0	8.5	18.8	0.4	4.0
Cheese	5.0	4.9	3.0	6.6	0.1	1.2
All meat	28.0	27.0	11.5	25.3	2.0	17.2
Butter	12.3	11.9	7.3	16.0	0.3	3.0
Margarine	13.5	13.1	4.5	9.9	2.6	23.1
Other fats	12.1	11.7	4.3	9.4	2.3	20.3
Biscuits	4.5	4.4	2.3	5.0	0.5	4.4
Total of items	89.9	87.0	41.4	91.0	8.2	73.2

The report recommends the use of (1) skimmed milk and fat reduced cheese; (2) poultry and non-fatty fish as an alternative to meat and meat products; (3) a movement towards the polyundsaturated margarines and cooking oils. All the recommendations would reduce the total fat and increase the P/S ratio.

The increase in dietary fibre can be achieved by the use of wholemeal flour and an increase in the consumption of fresh fruits and vegetables. The increase in dietary fibre would also provide additional health benefits in the form of a relief from constipation, a relief from diverticular disease and the retention of water in the intestines which dilutes and removes more quickly potentially harmful materials. Table 4.12 shows the dietary fibre content of some common foods.

Table 4.12 *Dietary fibre content of a selection of foods*

Food	% Dietary fibre	Food	% Dietary fibre
Bran	44.0	Red Kidney beans	
All Bran	28.0	(canned)	4.2
Cornflakes	1.5	Carrots	3.7
Coconut desiccated	23.5	Pasta wholemeal	
Almonds	14.3	boiled	3.3
Weetabix	12.7	White flour	3.1
Rice Krispies	0.9	White bread	2.7
Shredded Wheat	12.3	Unpolished rice	
Wholemeal flour	9.5	boiled	1.5
Wholemeal bread	8.5	Special K	1.2
Apples	2.0	Pasta white boiled	1.0
Peanuts	8.1	Polished rice boiled	0.8
Baked beans	7.3		
Broad beans	4.2		

EXERCISES

1 What is an emulsion and how are emulsions made? What influence do you think that temperature would have on the stability of the emulsion?

2 Devise an investigation in the field of catering that you could carry out relating to the amount of fat in a particular commodity. Discuss the reasons why you chose that particular topic.

3 Experiments show that vegetable oils are more resistant to oxidative rancidity than animal fats. This is surprising since vegetable oils have higher iodine values.

Explain (a)   What is oxidative rancidity?
        (b)   What is the iodine value of a fat?
        (c)   Why are vegetable oils more resistant than animal fats?

4 In recent years pressure fryers have been introduced into food outlets. What advantages if any do they have over the normal fryer?

5 Describe how the sucrose content of foods can be reduced. How would you devise experiments to see if sucrose reduced products were acceptable to your customers?

6 Explain the meaning of the term 'colloid' and discuss the problems of forming gels with starch and pectin.

7 Describe the role of proteins in foods and how can an understanding of their chemical properties enable you to make successful products?

8 If a meal should provide one-third of the RDA of a 35 year old moderately active male, how do MEALS 1 and 2 compare?

9 Try and find out by writing to some large catering organisations if they have a healthy eating policy – some do such as Compass Services 'Health line'.

10 Write an account on nutritional deficiencies that are currently of concern, eg mineral difficiency, highlighting the groups of people who are at 'most risk' and suggesting how they can improve upon their nutritional status.

5

Micro-organisms, Structure and Growth

The caterer is concerned with the activities of micro-organisms because these organisms are involved in food spoilage and decay and may be the cause of food poisoning and food-borne disease.

Micro-organisms may be divided into five groups according to characteristics of structure and nutrition:

1 PROTOZOA

These are single-celled animals. They are dependent for their nutrition on other organisms which they ingest or on organic materials made by other living creatures. Some protozoa are parasitic and invade the living cells of plants and animals. An example of such a parasite is Entamoeba histolytica which causes amoebic dysentary in man by invading the intestines, if swallowed in contaminated water or food.

2 ALGAE

These are simple plants, many are small microscopic structures but some may become quite large, for example seaweeds, where fronds may reach two metres in length. Algae are true plants and obtain their energy from sunlight by photosynthesis. Some algae are green in colour others may be brown or bluish-green. Algae are generally aquatic and may grow in stagnant ponds and water tanks but some may grow on damp north facing walls. The group is of no great significance to caterers.

3 FUNGI

This group includes moulds and yeasts. They are simple plants but they are not pigmented so they cannot gain energy from sunlight by photosynthesis. They are therefore either *saprophytic*, obtaining nutrients from the dead and decaying tissues of other organisms or *parasitic*, obtaining nutrients by invading the tissues of another living organism which acts as host.

Many fungi are microscopic but some may grow to structures as large as dinner plates.

Moulds and yeasts are important agents of food spoilage. Yeasts are important in brewing and baking and some moulds play roles in the development of flavour in such foods as cheese and salami. Some fungi, such as mushrooms, are eaten as food.

5.1 Cells of micro-organisms

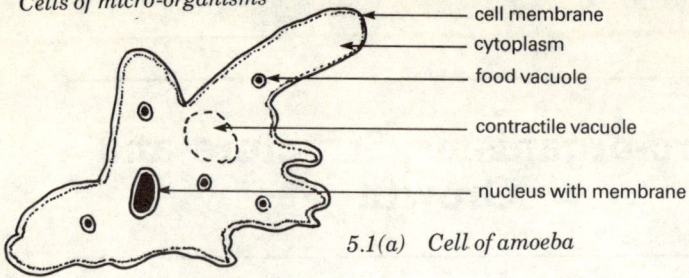

cell membrane
cytoplasm
food vacuole
contractile vacuole
nucleus with membrane

5.1(a) Cell of amoeba

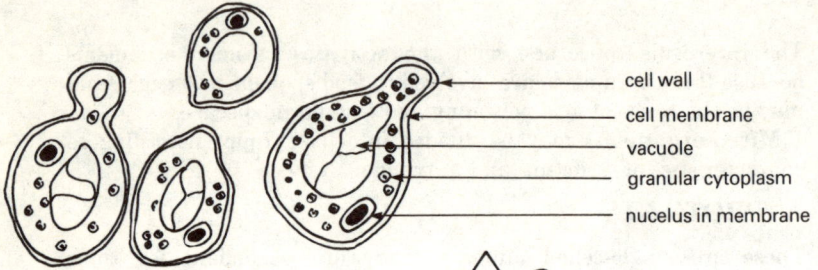

cell wall
cell membrane
vacuole
granular cytoplasm
nucelus in membrane

5.1(c) Yeast cells

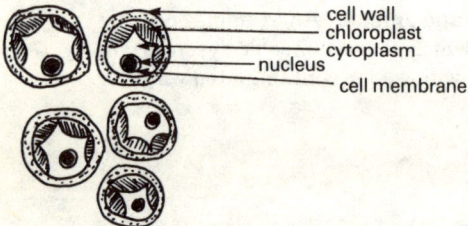

cell wall
chloroplast
cytoplasm
nucleus
cell membrane

5.1(b) Algae: cells of pleurococcus

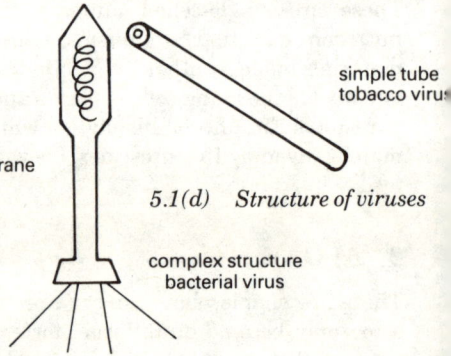

simple tube
tobacco virus

5.1(d) Structure of viruses

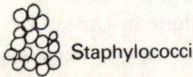

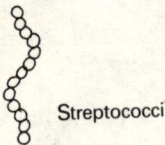

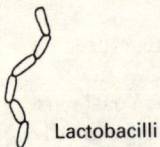

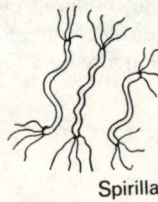

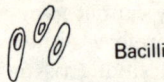

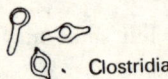

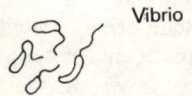

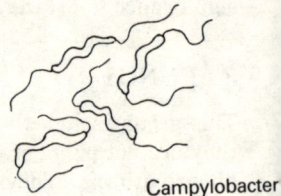

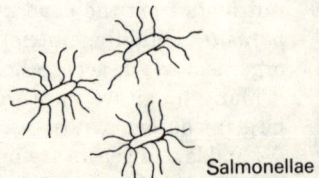

complex structure
bacterial virus

Staphylococci

Streptococci

Spirilla

Bacilli

Vibrio

Campylobacter

Clostridia

Lactobacilli

Salmonellae

5.1(e) Types of bacterial cells

4 BACTERIA

These are unicellular microscopic organisms. In size and shape they range from spheres 1 μm in diameter to thin rods up to 10 μm in length. They do not contain the green pigment chlorophyll and thus, similar to the fungi, bacteria are either *saprophytic* or *parasitic* although a very small sub-group can utilize sunlight energy.

Bacteria are active in the decay of plant and animal remains and are major agents in food spoilage. Parasitic species may also cause disease in plants, animals and man. Some of these pathogenic, disease-causing bacteria may be spread by contaminated food and water.

Caterers are concerned with the production of food of good quality and so must control the activity of bacteria to prevent food spoilage and the spread of food-borne disease.

5 VIRUSES

These are strange life forms being neither plants nor animals, they are strict parasites, they have no enzyme systems and cannot live or reproduce outside the body cells of their particular host. They are all ultra-microscopic, 20 mμ to 300 mμ in size, thus very much smaller than bacteria. Some are simple in structure, tubes and spheres, others are more complex syringe-like structures. They all contain nucleic acids. When a virus enters its host cell these nucleic acids take over the activities of the cell which becomes a factory to make more of the virus. Viruses are the cause of many diseases as they invade the cells of man, animals, plants and even bacteria. Some virus diseases of man can be spread by contaminated food which acts as a vehicle by which the virus may be passed from one host to another. Some viruses may cause gastro-enteritis with symptoms of vomiting and diarrhoea similar to food poisoning.

In the case of the disease AIDS, the virus is present in the body fluids of its victims and infection is spread by contact with such body fluids as blood, lymph, semen and breast milk. The virus may enter the body by means of infected hypodermic needles or through cuts and abrasions in the skin and mucosa lining body orifices.

Of the five groups of micro-organism **bacteria** and **fungi** are of outstanding importance to the caterer.

FORM AND STRUCTURE OF FUNGI

This is a large and complex group, its members important to caterers in three ways: *moulds*, which cause food spoilage or unsightly patches on damp walls; *yeasts*, used in fermentation processes in brewing and baking, and large edible *fungi*, such as mushrooms, used in cooking.

Colonies of yeasts and food spoilage moulds can be seen with the naked eye but the details of such growths can only be seen with the microscope.

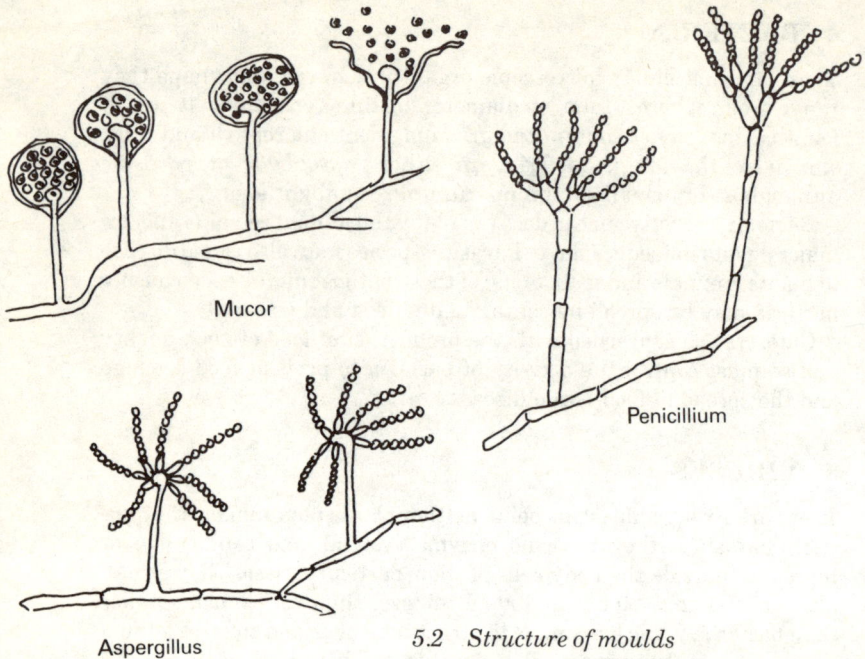

5.2   *Structure of moulds*

Yeasts

Yeast cells can be almost spherical to spindle shaped, from 5 μm to 15 μm in diameter. There is a non-living outer wall of polysaccharide which protects the living structures within. There is a cell membrane which closely follows the inner surface of the cell wall, then a jelly-like mass of granular cytoplasm; this granularity is due to the presence of oil droplets and glycogen granules. Floating in the cytoplasm is the nucleus, a spherical body containing genetic material enclosed in a membrane. Also present are two or three fluid-filled sacs enclosed with membranes, these are called *vacuoles*. Reproduction may occur by simple budding where a daughter cell grows out like a blister from the parent cell, but sexual reproduction can also occur with interchange of genetic materials between two cells and the formation of ascospores.

Moulds   Figure 5.2

With the naked eye mould growths appear as tiny mats of cotton wool, each a mass of minute inter-lacing threads perhaps pigmented black, yellow or green. The whole mass is known as the *mycelium*, the threads are called *hyphae*. Hyphae may be made of columns of cells joined end to end but in some moulds the cross walls are lost so that one cell is contiguous with the next forming a long hollow tube. In structure the mould cell is much like that of yeast with cell wall, cell membrane, cytoplasm, nucleus and vacuole. Reproduction is by the production of spores by special structures at the ends of some of the hyphae. Different species of moulds can be identified by the form of these spore-forming structures. Some moulds also use a sexual process where hyphae from two neighbouring colonies fuse together to form a *zygospore*. Different types of sporing structures can be seen in the common moulds Mucor, Penicillium and Aspergillus.

Mucor   Figure 5.2(a)

In Mucor each of the spore-producing hyphae ends in a globular sac, the sporangium, inside which the black pigmented spores are produced. When the spores are ripe the wall of the sporangium splits open to release a cloud of spores into the air by which they are carried until they fall onto the surface of a suitable food source or substrate. When on a suitable substrate mould spores will germinate, hyphae growing out of the spore case to produce a new mycelium visible to the naked eye in three or four days.

Penicillium   Figure 5.2(b)

Moulds of genus *Pencillium* have become well known through the antibiotic penicillin produced by species Penicillium notatum. However, other species are involved in food spoilage and others, such as Penicillium roqueforti, are involved in the ripening of blue cheeses. Penicillium moulds do not produce spores in a sporangium. Hyphae which are to produce spores terminate in a group of special bottle-shaped cells called *sterigmata* which appear like fingers on the hand of a skeleton. Blue-green spores, called *conidia*, are budded off, naked, from the necks of these bottle-shaped sterigmata and hang together as short beadlike strings. Ripe spores break off one at a time to float away in passing air currents.

Aspergillus   Figure 5.2(c)

This genus also produces naked spores. Cells which are to bear spores end in knob-like vesicles on the surface of which stand numbers of sterigmata each with a short chain of spores (*conidia*). Spores are generally pigmented brown or amber-yellow. Spore heads when magnified under the microscope resemble the seed heads of the dandelion.

It is important to note that moulds can produce large numbers of spores which are then released into the atmosphere. The air around us must be regarded as loaded with mould spores. Mouldy food must be discarded as soon as it is found and wrapped in paper or a polythene bag to contain the spores and thus prevent the spread of contamination.

CLASSIFICATION OF BACTERIA

Bacteria are first divided into groups according to cell shape. Some organisms have cells which are spherical or nearly spherical; some have rod-like cells, others are long thin cells twisted in the form of a coil or helix.

Spherical bacterial cells are called *cocci* (from the Greek word meaning a berry), *Coccus* = a single cell, *cocci* = a number of cells. Some cocci hang together in irregular clumps similar to bunches of grapes, these are called *Staphylococci* (Greek, *staphyli* = grapes). Some cocci hang together in chain-like formations, these are called *Streptococci* (Greek, streptus = chain).

Some rods are called *bacilli* (Latin, *bacillus* = little rod:plural *bacilli*). Some rods hang together in chains but because they are generally associated with milk and milk products are called *Lactobacilli* (Latin, *lacteus* = milky). Some bacteria are motile, that is they can swim through water by whip-like structures called *flagella* (*flagellum*:singular).

The structure of the bacterial cell is essentially simple (figure 5.3). There is a non-living cell wall generally covered on the outside with a protective capsule of gummy material either polysaccharide or polypeptide in nature. The cell wall can be either largely of protein or largely of complex carbohydrate material, *hexosamines*. These differences in composition can be demonstrated by use of a staining technique developed by Gram, an early microbiologist, in 1880.

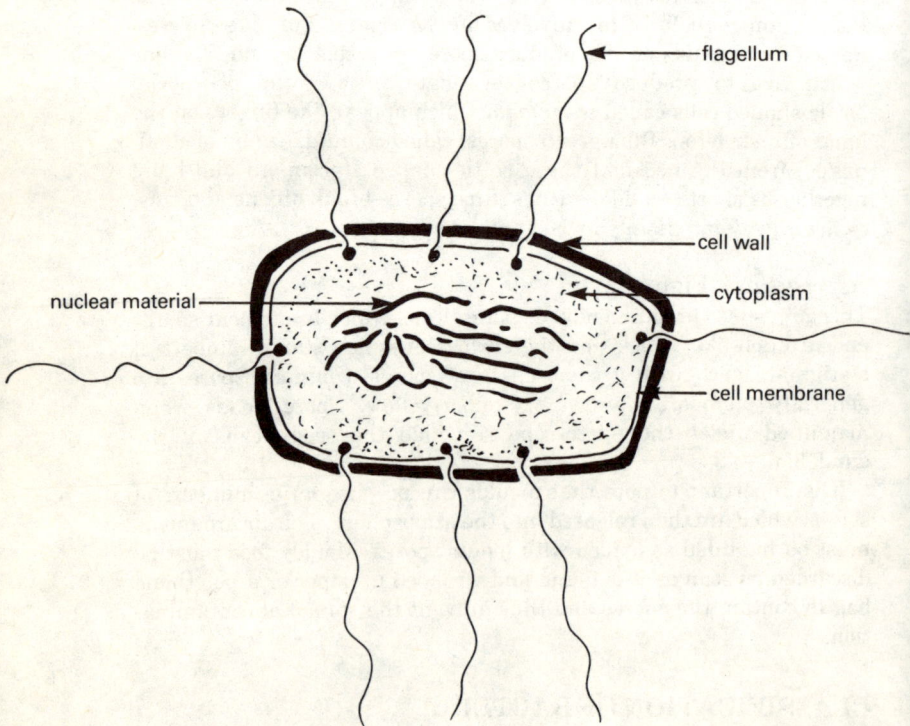

5.3   Structure of bacterial cell

Stained cells, which appear under the microscope a violet colour, have a cell wall containing hexosamines and are said to be *Gram positive* (+), those cells having a red colour have cell walls containing protein and are termed Gram negative (−). It is thus possible to have Gram positive cocci, Gram negative cocci, Gram positive rods and Gram negative rods.

Inside the cell wall the living cell is contained in a membrane which follows closely the inner surface of the cell wall. The mass of the cell is made up of a jelly-like substance (like egg white), the cytoplasm. Suspended in the cytoplasm are threads of genetic material DNA in a diffuse mass not enclosed in a membrane.

Some groups of bacteria have the ability to form spores, minute structures about 1 μm in diameter which resemble dried peas. These structures are resistant to heat and desiccation. By forming spores these bacteria may survive cooking or processes of food preservation. When suitable conditions return, spores will germinate to give active vegetative cells. Spore formation is a means of survival it is not a process of asexual reproduction as is the case with moulds and fungi.

REPRODUCTION IN BACTERIA

Reproduction in bacteria is by cell division, an active vegetative cell may divide into two daughter cells every 15 to 20 minutes if conditions are satisfactory. In a few hours very large numbers may be formed, numbers doubling every 20 minutes or so.

Time	0	20 mins	40 mins	60 mins	2 hours	3 hours
Cell numbers	1	2	4	8	64	512

Reproducing at this rate bacteria could swamp the world but as numbers increase so conditions become less satisfactory and so reproduction declines.

The growth of the colony   Figure 5.5

Consider what may happen to the growth of *Streptococcus lactis* in a bottle of milk left neglected for some days. Initially, for the first hour or two, no changes would occur whilst the cells explore the medium and prepare the necessary enzyme systems to exploit the nutrients available. This stage is known as the *lag phase*. A number of factors will affect the length of the phase such as the nature of the cells and their recent history, the nature of the medium and the temperature.

Feeding on the milk bacteria will begin to grow and reproduce, to produce many thousands of cells in a few hours. This stage is known as the *log phase* or *exponential phase*. However, as the population in the bottle of milk increases so the concentration of available nutrients, such as lactose, will fall, similarly, oxygen dissolved in the milk will be used up. Lactose will be fermented to lactic acid so the milk will become more and more acid and less and less favourable for microbiological growth.

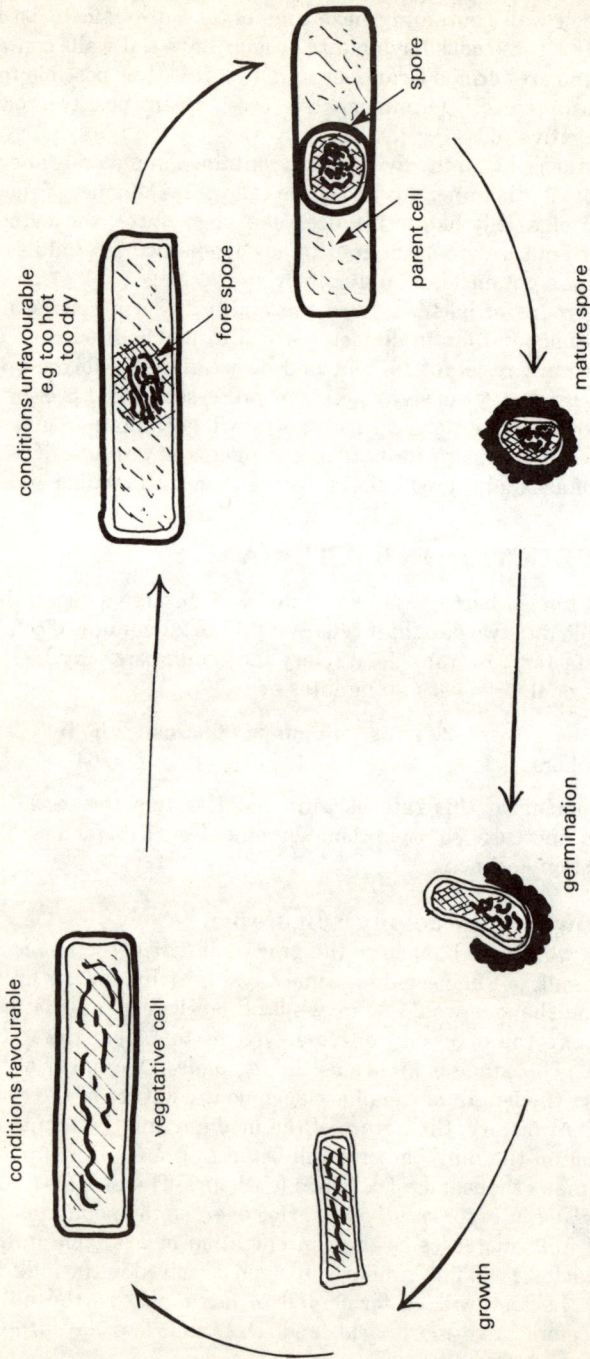

5.4   The spore cycle

A stage will be reached when, although reproduction still occurs, cells begin to die and there is no overall increase in numbers of living cells so that numbers remain high but constant, this is the *stationary phase*. Gradually the rate of cell deaths increases and the rate of reproduction falls so the numbers of living cells fall steadily and the colony enters the *decline phrase*. A time may be reached when all bacterial activity ceases. The milk by this time would be undrinkable. In the production of yoghurt a cycle of events such as described would occur.

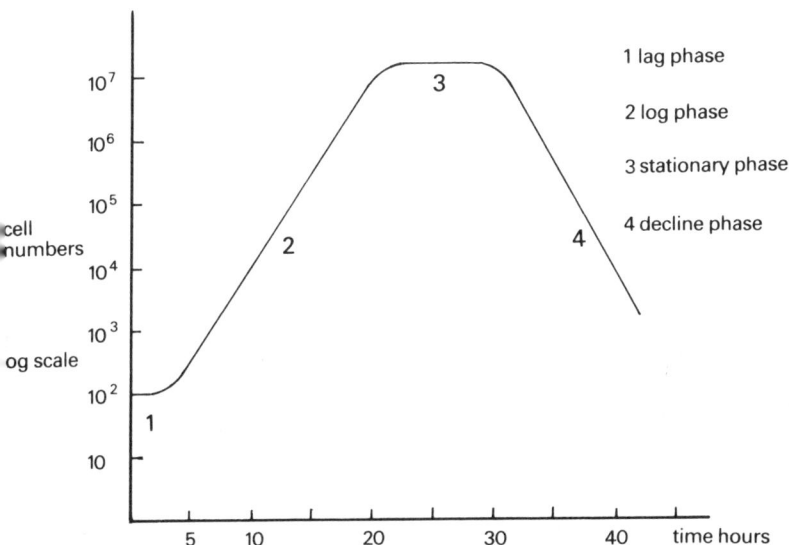

5.5 The growth of the colony

The nature of the cycle and the shape of the growth curve is affected by temperature. Refrigeration would extend the lag phase, slow down the rate of reproduction and reduce the slope of the log phase. Addition of further nutrients when the colony was approaching the stationary phase, such as mixing fresh milk with old, would extend the log phase and produce more active cells.

This cycle of events occurs wherever organisms colonise fresh material. Indeed man himself is on such a cycle in his colonisation and exploitation of the planet Earth. As man exhausts Earth's resources and pollutes its environment his civilisation must go into evential and inevitable decline.

Factors governing the growth of micro-organisms
It has already been stated that the growth of micro-organisms will be affected by a number of environmental factors. These are:

1 Availability of nutrients.
2 Availability of water.
3 Availability of oxygen.

4   Temperature of environment.
5   Acidity/alkalinity of medium. Hydrogen ion concentration.
6   Presence of inhibitory substances.

1   Availability of nutrients

Micro-organisms such as bacteria and moulds require food materials for two purposes:

(i) as sources of energy, generally carbon compounds, carbohydrates like sugars, and organic acids;

(ii) as structural materials for the synthesis of protoplasm, cell walls, membranes and enzymes. This need is met by proteins and amino acids.

However, bacteria vary widely in their nutritional requirements. Many bacteria in soil and water can live on simple organic acids as sources of energy and use ammonia or even atmospheric nitrogen as raw materials for the synthesis of proteins. Micro-organisms involved in food spoilage require more complex organic substances, proteins, amino acids and vitamins. Some bacteria, for example Lactobacilli, are particularly fastidious and will not grow unless specific amino acids or vitamins are present. The perishability of a food item depends on the range of nutrients present. If the range of nutrients is wide, the range of micro-organisms the food will support will be wide also and the food will perish quickly. If the range of nutrients in a food is limited, then the range of micro-organisms it will support will also be limited and the food will be less perishable.

2   Availability of water

Life, even in its simplest form, cannot be sustained without free water, ie not chemically bound and free to move within a structure as in meat or vegetable or fruit tissue. Micro-organisms will vary in their requirements for water, bacteria having a greater need than moulds and fungi. The availability of water in food will govern the activity of micro-organisms present and thus the rate of decay. It is not the amount of water present that is important but its availability. The availability of water in foods is reduced by dissolved substances particularly sugars and salts. The availability of water is expressed in terms of water activity, a_w. The a_w of pure water is 1, that of milk, meat and fish being about 0.995 so that these commodities are highly perishable and will support the growth of a wide range of micro-organisms. Preserved foods such as cheese, bacon and kippers with a_w about 0.88 are much less perishable although will suffer attack from less demanding micro-organisms, particularly yeasts and moulds. Foods with a high sugar content such as jam and syrups with a_w between 0.80 – 0.75 will support the growth of very resistant moulds and yeasts only. Micro-organisms will not grow on foods with a_w below 0.60.

Table 5.1 *Availability of water and growth of micro-organisms*

Micro-organism	Minimum a_w for growth
Normal bacteria	0.94
Normal yeasts	0.88
Normal moulds	0.80
Salt tolerant (halophilic) bacteria	0.75
Sugar tolerant (xerophilic) moulds	0.65
Sugar tolerant (osmophilic) yeasts	0.60

3 Availability of oxygen

Micro-organisms have a range of requirements for oxygen, those which are dependent upon free oxygen in the air are called *aerobes*. Aerobes grow best in a normal atmosphere of about 24%–15% oxygen. There are some micro-organisms however which grow only in the absence of oxygen, these are called *anaerobes*. A third group grow best in an oxygen concentration below that of the normal atmosphere between 15%–5% oxygen these are said to be *microaerophilic*. Some micro-organisms, whilst growing better in the presence of oxygen, can grow in its absence. Such organisms are termed *facultative anaerobes*.

4 Temperature of environment

In the living world there are those plants and animals which inhabit tropical regions, others prefer sub-tropical temperatures and others thrive in temperate regions. Similarly micro-organisms can be divided into three groups according to the temperatures at which growth occurs. The three groups are termed (1) *psychrophilic*; micro-organisms growing in cold conditions; (2) *thermophilic*; those organisms which grow in hot conditions; and (3) *mesophilic*; those organisms which grow at moderate temperatures. Within a range of temperatures there will be a point at which an organism is most active, the optimum temperature, either side of this point activity will steadily decline until it ceases altogether.

Table 5.2 *Temperature and growth of micro-organisms*

Group	Activity temperature range °C		
	minimum	optimum	maximum
Cold tolerant psychrophilic	−10	10	25
Middle group mesophilic	10	35	45
Heat tolerant thermophilic	40	55	95
Moulds	5	25	40

Moulds involved in food spoilage do not fall neatly into the *mesophilic* group, these organisms are active at slightly lower temperatures than mesophilic bacteria. *Psychrophilic* bacteria are found in the soil and in sea mud hence contaminate fish, particularly demersal species. The marked perishability of fish is due in part to this contamination. *Thermophilic* bacteria also are found in the soil and in hot springs and geysers. Their importance to caterers is that they can cause souring in canned goods and stock-pots. *Mesophilic* bacteria are associated with man and warm blooded animals and are to be found on body surfaces, in tracts and orifices, and in the habitat in general. The group contains not only food spoilage types but also pathogenic bacteria involved in food-borne disease and food poisoning.

5   Hydrogen ion concentration

Gardeners know that the acid/alkalinity of the soil affects how plants will grow in it. Similarly the growth of micro-organisms in food will be affected by the hydrogen ion concentration of the item.

In general most bacteria are neutrophilic and grow best between pH 6.0 and 8.0. Some are more tolerant and are active over a wider range between pH 4.0 and 9.0. Moulds and yeasts are more tolerant of acidic conditions and are active between pH 2.0 and 9.0.

There is only one common food item which is alkaline, that is egg, particularly egg white which is about pH 9.0. It is good practice to make food dishes as acid as is acceptable to control the activity of micro-organisms.

6   Presence of inhibitory substances

Microbiological activity can be affected by the presence of chemical substances. Some of these substances can be simple compounds such as sulphur dioxide, nitrites, or organic acids, others are complex organic substances such as antibiotics. All these substances affect the metabolic enzyme systems of micro-organisms and check their growth and development. Some of them have value as preservatives.

IMPORTANT GROUPS IN FOOD BACTERIOLOGY

1 Spore-forming bacteria

A knowledge of spore-forming bacteria is important to the caterer in that their spores are able to survive processes of boiling, baking and roasting. In cooling food, spores may then germinate, colonies grow and spoilage or perhaps food poisoning may result. Amongst spore-forming groups two genera are of note, genus *Clostridium* and genus *Bacillus*. The two genera have some features in common. They are widely distributed in the soil. They are Gram positive rods, they have hexosamine type cell walls, members ferment sugars to give acids thus cause souring.

Genus Bacillus

Members of this group are aerobic and grow best in air but some are facultative anaerobes. Spores form inside the parent cell but do not distend it.

Bacillus subtilis The organism is ubiquitous in the soil and in cereal products. It produces a range of food-splitting enzymes, amylase and protease and can ferment most sugars except lactose. Because its spores survive baking, the organism can cause a defect in bread known as *rope*. As the spores germinate and cells grow in the bread, enzymes break down the crumb to a soft sticky khaki-coloured mass with a sour fusty smell. The problem is more common in summer when hot bread takes longer to cool. The organism is mesophilic and grows between 10 and 50°C. Spoilage is prevented by use of proprionic acid as a preservative.

Bacillus cereus This organism is also found in the soil, in cereals and in milk. In its range of enzymes are amylase, protease and also lecithinase. This enzyme can cause spoilage in milk. The lecithinase removes the lecithin envelope around the cream droplets in the milk which then clump together. Should the milk be used in tea or coffee these lumps of cream float on the surface and look unsightly although the milk will taste well enough. This defect is known as *bitty cream*. Further changes can follow and the milk may curdle but, because Bacillus cereus does not ferment lactose, the milk will not taste sour, hence this type of spoilage is known as *sweet-curdling*. Bacillus cereus may also produce enterotoxins when growing in some food dishes, such as parboiled rice or flour thickened sauces, and thus this organism is one of those causing food poisoning.

Bacillus stearothermophilus This organism is a true thermophile and grows best between 50 and 75°C. The spores are more resistant to heat than is general for the genus and may survive, for example, the canning process. If spores survive in low acid canned vegetables such as peas, beans and carrots, the spores may germinate in the hot can after processing and the active vegetative cells ferment such sugars as are present to acids. This fermentation does not involve the production of gases, as does yeast fermentation for example, thus there are no signs of fermentation when cans are opened but the contents taste sour, hence cans which show this type of defect are said to be *flat sours*.

Bacillus anthracis This organism is similar to Bacillus cereus. It is found in the soil particularly in farmyards. It is a more serious pathogen and is parasitic in man and other mammals. It causes the disease anthrax. In man the disease is usually seen as large eroding ulcers on the skin where the organism has entered through small cuts and abrasions. Sources of infection could be farmyard soil, hides and fleeces, and meat. It should be stated that Anthrax is rare but it is unwise to ignore cuts received when handling meat and animal products. Anthrax does respond to treatment with antibiotics but deaths from the disease are still reported from time to time.

Genus Clostridium

This group of Gram positive rods are anaerobic. They are found in the soil, and do not grow in the presence of oxygen. They produce spores which distend the parent cell. Most members ferment sugars to give acids and gases, carbon dioxide and hydrogen. Some members are strongly proteolytic.

Clostridium perfringens (welchii) Clostridium perfringens is a common organism in the soil, it is also found in the intestines of herbivorous mammals and hence in their faeces, therefore ubiquitous in the farmyard. It is found on meat and soil-covered vegetables. It is strongly proteolytic and can cause putrifaction in meat and sterilized milk. However, it is pathogenic in man and causes disease in two ways:

1 If spores enter wounds, abrasions and burns, a serious infection can result in which, as the tissues are broken down by the invading organism, gases are given off, the result of enzyme activity, thus the name for this condition, *gas gangrene*. Mention has already been made of the need of care with wounds received when cutting meat.
2 Gastroenteritis may result from eating food containing large numbers of vegetative cells, so this organism is one of those concerned in food poisoning. Food poisoning from *Clostridium perfringens* in most victims is fairly mild lasting two or three days but deaths do sometimes occur.

Clostridium botulinum To the caterer this organism is potentially the most deadly. It causes the disease botulism which is fatal in three out of four cases. Clostridium botulinum is found in the soil, in sea mud, in the intestines of hebivorous mammals and thus in meat, fish and soil-covered vegetables. Should spores survive in canned meat, fish or low acid vegetables or semi-preserved items such as pâté, ham and sausage, then neurotoxins could be formed in the food by vegetative cells, which toxins are amongst the most virulent known. However, by strict control of thermal processing and the use of additives, food manufacturers ensure the safety of their products so that botulism is rare.

Clostridium thermosaccharolyticum The majority of Clostridia are not pathogens. Clostridium thermosaccharolyticum is an example of one involved in the spoilage of canned foods. It is found in the soil and as a contaminant on food crops. Its spores are heat resistant and may survive thermal processing, the organism is also thermophilic. Thus, as with Bacillus stearothermophilus, spores may germinate in hot cans after processing and active vegetative cells ferment sugars present not only to acids but also produce fair quantities of gas. This gas will distend the lids and bases of contaminated cans which firmly resist finger pressure to push them back. Hence the name for this type of spoilage, *hard swells*.

2   The enteric group, bacteria of the intestines

Bacteria exploit the habitat of the mammalian gut, some are *saprophytic*, harmless, whose presence goes unnoticed, others are *pathogenic*, whose invasion results in disease. All members of the group have features in common, they are Gram negative and have protein type cell walls, they do not form spores, all are mesophilic and ferment sugars to acids. The majority are motile but some important members are not.

Genus Escherichia

Members of this genus inhabit the intestines of mammals. Escherichia coli type I is found in man. There are a number of different strains of the organism so that invasion of the intestines by a foreign strain, for example when on holiday abroad, may result in mild gastro-enteritis. Similarly with babies under six months old, invasion of the intestines by Escherichia coli may result in gastro-enteritis which could prove fatal.

Escherichia coli are excreted in large numbers in faeces and are found in sewage. The presence of Escherichia coli type I in food indicates contamination with sewage and if found in the work place indicates very poor standards of personal hygiene. Environmental health officers and food microbiologists use the presence of the organism in food samples and swabs as an index of hygiene and food safety.

Genus Salmonella

This is a widespread genus of intestinal parasites found in the intestines of warm blooded animals, birds and mammals. They are excreted in faeces and are found in sewage contaminated water, so seafood taken from such water, particularly shellfish, may also be contaminated. There are some 1500 species, all pathogenic, of different degrees of virulence, some causing enteric fever others gastro-enteritis. Some are named after the associate disease, eg *Salmonella typhi*; some after renowned microbiologists, eg *Salmonella virchow*; others after towns with which they were first associated, eg *Salmonella montevideo*. Similar to Escherichia coli in form and habitat, one important difference is that whereas Escherichia coli ferment lactose Salmonellae do not.

Genus Shigella

This is a small but important genus, like the Salmonellae they are pathogenic intestinal parasites, they too do not ferment lactose but unlike the Salmonellae they are not motile. They cause dysentery. As with other enteric microorganisms they are excreted in faeces from infected victims and spread by sewage contaminated food and water. Two principal species are: *Shigella sonnei* and *Shigella flexner*.

3   The lactic bacteria

This group includes both spheres and rod-shaped Gram positive cells. As cells divide, division is always in the same plane and cells tend to hang together in chains. They are mesophilic and microaerophilic and are associated with milk and milk products. They ferment sugars to lactic acid and in some cases produce other substances as well, such as acetic acid, ethyl alcohol and carbon dioxide.

Genus Lactobacillus

This genus contains the rod forms. Some members are used in food manufacture. *Lactobacillus bulgaricus* is used in the manufacture of yoghurt. *Lactobacillus plantarum* is used in the manufacture of sauerkraut and pickled cucumbers.

Some species are involved in the production of silage. Acids produced from the fermentation of molasses preserves the silage from putrifaction. Lactobacilli may be involved in food spoilage and may cause souring in milk, wine, beer and fruit juices.

Genus Streptococcus

This genus contains Gram positive cocci which occur in pairs or chains and ferment sugars to lactic acid. They are to be found associated with man and other mammals, some saprophytic others pathogenic. The genus may be divided into four sub-groups.

(i) Pyogenic group   This group contains the pathogenic members such as *Streptococcus pyogenes*. This organism causes septic sore throats and scarlet fever. The organism can grow in milk and milk dishes thus these foods can act as vehicles for the spread of disease. Septic sore throat, for example, can be spread by an infected food handler coughing over a pan of custard.

(ii) Viridans group   Members of this group are saprophytes found in the mouth and on the skin of man and mammals, normally they are harmless but some such as *Streptococcus viridans* may cause disease. *Streptococcus viridans* present in the mouth may enter the body after tooth extraction and set up an infection in the heart which can prove fatal or leave the heart permanently impaired. Members of this group are also implicated in tooth decay because they ferment sugars to produce acids.

(iii) The lactic group   This group contains the important milk bacteria such as *Streptococcus lactis* and *Streptococcus creamoris*. They are found in milk and milk products, in cattle and in dairy equipment. They are the principal cause of the souring of milk. Strains of these organisms are used as cultures in the controlled souring of milk, in the manufacture of cheese, and to develop a butter-like flavour in margarine.

(iv) The enteric group   Some species of Streptococci are found in the intestines of man and other mammals along-side Escherichia coli, for example *Streptococcus faecalis* and *Streptococcus faecium*. As a group they can grow over a wide temperature range from 5 to 50°C. They have a low a_w (0.85) and can grow in salted foods such as ham, bacon and cheese. They are amongst the few bacterial species able to grow in alkaline conditions as high as pH 9.6. Although they do not form spores, cells are heat resistant and may survive in pasteurized milk and large canned hams and are potential spoilage organisms. As their habitat is the intestines and they are found in faeces, their presence in food and water indicates sewage contamination.

4   Bacteria of the skin: the micrococci

Micrococci are the principal organisms of the skin although other groups may be present. The organisms are sloughed off with skin scales and are thus found in house dust. They are transmitted to food during handling and preparation. Cells are spherical about 1 µm in diameter, Gram positive, and occur in irregular clumps. They are mesophilic and aerobic some microaerophilic. Group members are generally harmless saprophytes but there is one important pathogenic species Staphylococcus aureus.

Staphylococcus aureus

This organism is found on the skin and in the nose and throat of man. About one third of the community carry the organism as a permanent resident in their noses. The organism is pathogenic and causes septic skin conditions from pimples to boils, carbuncles and septic wounds of all types. The organism produces a range of enzymes to enable it to establish itself in a defect in the skin, protease to break down tissues, haemolysin to destroy red blood cells, leucocidin to destroy white blood cells and coagulase which coagulates blood plasma and so to seal off the site of infection from the body's defences. Together with all this, Staphylococcus aureus can grow in most food stuffs with a_w above a minimum of 0.86 and produces heat resistant enterotoxins which, when the food is eaten, cause food poisoning. Staphylococcus aureus is then another organism of importance to the caterer.

5   Proteolytic bacteria

This is a diverse group which causes spoilage in protein food such as slime and off-odours in meat, rots in eggs and the putrifaction of fish. Members of the group are Gram negative motile rods, mesophilic or psychrophilic, aerobic and do not form spores. They are to be found in the soil and sea mud wherever rotting proteinaceous matter is to be found. Two genera are of importance: Proteus and Pseudomonas.

Genus Proteus

This group may best be represented by its most common member *Proteus vulgaris*. This organism is strongly proteolytic and forms hydrogen sulphide, (the smell of school boys' stink-bombs), from sulphur bearing amino-acids and also an evil smelling organic compound indole. In meat and fish it is a cause of slime, off-odours and bone taint. In eggs it is the cause of the rotten egg smell of hydrogen sulphide. As the gas is given off, it reacts with iron present in the yolk to give it a black-brown discolouration of ferric sulphide.

Genus Pseudomonas

Members of this genus are noted for the production of blue-green pigments during the spoilage of protein foods. When growing in an egg a blue-green colour is given to the white. In fish they are responsible for the fishy odour of trimethylamine. Some members, eg *Pseudomonas fluorescens* produce fluorescent compounds which produce a ghostly glow in rotting food debris. One pathogenic member, *Pseudomonas pyocyanea*, can grow in wounds to produce bluish tinted pus. Some members produce lipase and can cause racidity in butter and margarine, others such as *Pseudomonas fragi* produce fruity odours in cheeses.

EXERCISES

1   Make a list of preserved food products in use in your kitchen. For each product on your list show how the factors governing the growth of micro-organisms have been applied to prevent spoilage.

2   What types of micro-organisms could be expected to grow in cook/chill dishes left over-long in the refrigerator? What type of spoilage would these organisms cause?

3   Study the HCIMA Technical Brief (sheet 2) *Precaution against AIDS*. Prepare a list of procedures to be adopted in housekeeping and food service. Explain the significance of each in terms of microbiological principles.

6

Food Poisoning and Food Hygiene

Food poisoning may be defined as: sudden and severe gastro-enteritis following the consumption of tainted food. These poisoning taints may be classified according to their nature and origins:

1 Chemical poisons
2 Botanical poisons
3 Bacteriologial poisons.

1 CHEMICAL POISONS

Various chemical substances can cause vomiting and diarrhoea. Metals, antimony, arsenic, cadmium, lead and zinc can cause such trouble. It is important that care is taken in the purchase of equipment that it is fit for use with food. In the repair of equipment food grade solders and fluxes must be used. Food may become contaminated by careless handling of insecticides, rat poisons and weedkillers, and in turn may cause sickness or even death.

2 BOTANICAL POISONS

Innocuous looking vegetables may contain sufficient toxic substances to cause poisoning. Green potatoes contain the alkaloid poison solanine and are thus not fit for consumption and should be returned to the supplier. Rhubarb contains the poison oxalic acid. Red kidney beans contain poisonous haemagglutenins which are adequately destroyed only by vigorous boiling. Deaths occur every year from the consumption of poisonous fungi, for example eating the Death Cap (Amanita phalloides) in mistake for the Field Mushroom (Agaricus campestris). Caterers should use only mushrooms obtained from a reliable trade source. Some moulds produce poisonous mycotoxins when growing on food. Mouldy food should be seen as unfit for human consumption and discarded. It is not sufficient to 'cut out' the mould growth as the mycotoxins diffuse into the depth of the product.

3 BACTERIOLOGICAL POISONING

Food poisoning from bacteriological sources is by far the most important. Whereas numbers of victims of other types of poisoning are counted in tens in the United Kingdom, victims of bacteriological food poisoning are counted in tens of thousands. Poisoning from bacteria can arise in two ways:

1 Food may be infected with active pathogenic bacteria; – *food infections*.
2 Food may cause poisoning because bacteria have grown in it after harvesting and produced toxic substances which survive in the food to be ingested when the food is subsequently eaten: *toxin poisoning*.

Food infections
This type of poisoning is caused by the presence of active vegetative bacteria in a food item at the time it is put into the mouth. Species involved are:

Salmonella species   Clostridium perfringens
Campylobacter jejuni   Vibrio parahaemolyticus

Salmonella species
This must be regarded as a very successful genus because of the large number of species and, as parasites, they are widely distributed amongst all warm-blooded animals, both birds and mammals. Although infection often results in sickness and occasionally death of the host, this is not always so. Salmonellae, even the most virulent species, can set up a relationship with the host so that organisms inhabit the intestines but do not produce symptoms of gastro-enteritis. Such a relationship is termed *commensalism*. Individuals who are infected in this way are termed *carriers*. Carriers amongst a team of food-handlers can be a source of infection unless very high standards of personal hygiene are constantly maintained. Some species may pass through the intestinal wall into the blood stream then into the deep tissues. Salmonellae can enter hens' eggs before the shell is deposited, or be present in milk as it is secreted.

   In the kitchen, sources of Salmonellae are meat, offal and uncooked meat products, poultry and game, eggs and milk, fish, especially shell fish from sewage contaminated waters. (Now that caterers are required to use only pasteurized milk in the production of food for sale this item may be regarded as free from contamination.) Salmonellae do not form spores and are destroyed at temperatures of 65°C and above. Their death is certain in well-cooked meat and poultry dishes. They may survive in dishes were temperatures do not reach 65°C such as undercooked beef and egg based sauces such as sauce hollandaise. Another danger is the transfer of organisms from contaminated food to items which are not going to be cooked, for example from egg to cream by an unwashed whisk or from raw chicken to cold roast pork by a dirty knife or by a drip from meat onto triffle in a badly managed refrigerator. Salmonellae are mesophilic, they show little activity below 10°C and none below 5°C, however they can survive in frozen foods for 2 to 5 years and soon become active again when the product is defrosted. Care is required in the handling of frozen poultry where 25% or more of carcases should be considered contaminated. Defrosting should be done in the refrigerator or in cold water. Defrosting should be complete before cooking is begun to ensure all parts of the roasted fowl reach at least 63°C. Salmonellae cause two thirds of all cases of food poisoning yet they are easily controlled and readily destroyed by normal good kitchen practice.

Campylobacter jejuni

This organism, with others of the same genus, is a late arrival in the field of food hygiene; considered of little account before 1980, it now rivals the Salmonellae as a cause of gastro-enteritis. This is due inpart to improved methods of diagnosis and laboratory isolation and identification but changes in food policies, production methods and practices may also have played a part. Campylobacter jejuni is a highly motile, Gram negative spirillum, a short slightly twisted rod shaped cell about 3 µm × 0.5 µm with a single whip-like flagellum also about 3 µm in length at each end of the cell. It is mesophilic and microaerophilic. Campylobacter species have been isolated from a range of farm animals, poultry and other birds, rats and mice, domestic pets and primates such as monkeys. Although in humans those convalescent from enteritis may be carriers of the organism for a month or two, long term carriers are unknown. This intestinal parasite can be considered along with the Salmonellae and enters the kitchen in the same fresh and frozen foods; meat and meat products, poultry and game, eggs, unpasteurized milk and shell fish from sewage contaminated waters. Campylobacter is destroyed along with Salmonellae by pasteurization, so cooking to an internal temperature of 65°C effectively eliminates Campylobacter. The organism appears to be virulent, and relatively low numbers are required to cause infection, for example a small outbreak was caused when water from washing chicken carcases was splashed over prepared salad items. Campylobacter seeded onto the surface of raw food survive only two or three hours but in milk can survive for nearly a month. In general, measures to control and destroy Salmonellae will also control and destroy Campylobacter.

Clostridium perfringens

An account of this organism has been given in chapter 5. It is ubiquitous, an every day visitor to the kitchen on meat and soil-covered vegetables. Organisms form spores which survive normal methods of baking and roasting. To cause food poisoning very large numbers of several millions of active vegetative cells must be swallowed in food. In the hostile environment of the stomach, the cells form spores. The spore cycle takes some two hours when the mature spore is released, at the same time enterotoxins are secreted which in turn cause intestinal pain and diarrhoea. Foods responsible for outbreaks are meat and poultry dishes; stews, casseroles and pies which have been mishandled in some way. Usually the food has been allowed to cool too slowly, kept at kitchen temperatures for several hours without adequate refrigeration or kept hot without attention to temperature regulation so that the temperature of the food has fallen below 55°C. The organism is anaerobic and mesophilic but growth characteristic are unusual with minimum at 15°C, maximum at 53°C and optimum about 45°C. At the optimum the lag phase is almost absent and cell numbers double every eight minutes. Thus under these conditions active cells could increase 100,000 fold in two hours. Prevention of perfringens food poisoning lies in temperature control of food items during preparation and service, and these simple rules should be followed:

1   Adequate cooking to an internal temperature of 65°C.
2   If the product is to be held cold then it must be cooled quickly from 65 to 10°C in two hours maximum, then held in a refrigerator until required.
3   If the product is to be kept hot, the temperature of 65°C must be maintained throughout the food for the whole of the storage period.
4   Should cold dishes be reheated then this should be done quickly and all parts of the product brought to 65°C before service.

Vibrio parahaemolyticus

This is a very rare organism in the Western world but it has been the cause of many outbreaks in the Far East, in Japan and Thailand. The organism belongs to the same family as Campylobacter, the Spirilla-cae, and is similar in size and shape. Cells are Gram negative, small thick rounded rods about $2 \mu m \times 0.6 \mu m$ with a single whip like flagellum. The natural habitat of *Vibrio parahaemolyticus* is the sea, especially the coastal waters of Japan and Malaysia but the organism has also been found in European coastal waters. The organism is mesophilic and halophilic, it grows over a wide temperature range, between 3 and 44°C with an optimum at 35–36°C. It requires a salt concentration between 0.5 and 8.0% with an optimum at 2–3%. In the United Kingdom the few cases that have occured have involved shell fish, principally frozen prawns of Malaysian origin, although crab from British waters has also caused trouble. The symptoms are nausea, vomiting and severe diarrhoea similar to dysentery which may lead to dehydration. Onset is generally 16–18 hours after eating contaminated seafood, sickness lasts for two to five days but victims may require several weeks convalescence after a severe infection. The organism does not form spores and is destroyed by pasteurization at 65°C and above. Suspect foods such as frozen Malaysian prawns should be handled with care, defrosted in the refrigerator and kept cold. Water seeping from thawing seafoods should also be disposed of with care, so as to prevent cross-contamination. With crab, the normal practice of simmering in salted water for 40 minutes should free the product from infection.

Toxin poisoning

This type of poisoning is caused by the presence of toxins, which are poisonous substances produced in food items by bacteria growing in the food after harvesting during storage, preparation or service. The food may be sterile at the time of consumption but the toxins left active. For example, an outbreak of food poisoning occured in England from drinking reconstituted dried milk. The milk was dried in the USA the year before, the powder was free of active pathogenic bacteria but contained an enterotoxin formed in the milk by bacteria growing in the milk before it had been dried. Thus when the milk was used as food a year later it resulted in food poisoning.

Bacteria involved in toxin poisoning are:

> Staphylococcus aureus
> Bacillus cereus
> Clostridium botulinum.

Staphylococcus aureus

Details of this organism are given in chapter 5. Staphylococcus aureus, common on the skin and in the nose of man, will grow on any food material with an a_w above 0.86, in the temperature range 7 to 46°C. As the organism grows it secretes a powerful enterotoxin into the surrounding food, if the food is subsequently eaten then food poisoning will occur. The symptoms are nausea, abdominal pain, violent vomiting and diarrhoea, there is no evidence of fever; rather the body temperature may be a degree or two below normal. Onset is sudden between 1 to 6 hours after eating contaminated food. Complete recovery is normal in two or three days but death can occur in the very young or very old. The toxin is heat stable and is not destroyed by normal cooking methods, even pressure cooking for 10 to 15 minutes. These various factors are illustrated by an interesting outbreak following the consumption of hot potato chips. The chips were prepared by hand early in the day by a catering worker who had an open suppurating wound on the left hand which served as a source of Staphylococci. The chips were blanched in hot oil at 180°C for about two minutes then left on wooden trays in the kitchen until required later that evening. This gave the time and warm conditions required by the organism to grow in numbers and produce its toxin. When required for service the chips were refried at 180°C for a further 5 to 8 minutes. Hot oil at this temperature would kill the Staphylococci but would not destroy the enterotoxin they had produced which subsequently caused food poisoning in those who ate the chips. This case shows that Staphylococcus aureus will grow readily in the warmth of a kitchen, on ordinary food materials and produce heat stable enterotoxins which will survive subsequent cooking to high temperatures. The case also gives an example of a gross breach of hygiene regulations and failure by management to ensure their staff were fit to handle food.

Bacillus cereus

An account of this organism is given in chapter 5. Bacillus cereus causes two forms of food poisoning; there is a mild form of abdominal pain and diarrhoea 8–16 hours after eating contaminated food similar to perfringens poisoning; there is a more severe form with a speedy onset of 1–5 hours with symptoms of nausea and vomiting followed by diarrhoea similar to staphylococcal toxin poisoning. The mild form is associated with cereal dishes, wheat flour thickened soups, sauces and custard fillings. The more severe form found in the United Kingdom is associated with Chinese 'take away' rice. The usual story behind these outbreaks is that a batch of several pounds of parboiled rice is prepared well in advance of service and left as a warm moist mass at the back of the kitchen. When required for service, portions are taken for further cooking by boiling or frying. The spores of Bacillus cereus are commonly found on cereals and they survive normal methods of cooking. Thus in a mass of parboiled rice left in a warm kitchen these spores would germinate and the vegetative cells grow rapidly in numbers and produce enterotoxins. These enterotoxins are, like staphylococcal enterotoxins, highly stable and not destroyed by normal cooking methods and so survive in the reheated product. Similar cycles of events could occur with soups, sauces and custards. The growth of Bacillus cereus in cereal dishes is enhanced by the addition of proteins from meat, milk and egg.

Clostridium botulinum

Details of this organism are given in chapter 5. It is the cause of the disease botulism. Botulism is rare in the United Kingdom but there are some 50 cases each year in the USA and a similar number on the Continent of Europe. It is the most feared form of food poisoning because of its high mortality. The species is anaerobic, forms spores and there are six different types labelled A to F. The toxin produced when the organism grows in food is one of the most potent poisons known but it is not so stable to heat as other food poisoning toxins. The symptoms of botulism are different from those generally associated with food poisoning because the toxins produced by the various types of the organism are not enterotoxins but neurotoxins, they are absorbed into the blood stream and affect the central nervous system and not primarily the intestines. Thus the symptoms may suggest drunkenness, lassitude, dizziness, double vision, staggering, collapse, paralysis and finally death. Onset of the disease occurs within one to three days after consumption of tainted food and may take seven or eight days to reach its climax. Foods associated with botulism are: home canned or bottled low acid vegetables; semi-preserved meat products, hams, pâté, thick liver sausage; smoked fish and fish products. Cultivated mushrooms packed in air-tight film wrappings can support the growth of Clostridium botulinum resulting in toxin production in three or four days.

Organism	Sources	Foods involved	Nature of poisoning	Time of onset	Symptoms
Salmonellae 1500 species eg S typhimurium S virchow S montevideo	Intestinal parasite of warm blooded animals; man, other mammals and birds Aerobic	Meat and poultry dishes and pies Eggs and egg dishes Milk and milk dishes	*Food infection* Small numbers of living bacteria required 1000/g food	12 to 24 hours	Abdominal pain Frequent vomiting Diarrhoea Fever Dehydration Death <1%
Campylobacter jejuni	Intestinal parasite of warm blooded animals especially cattle and poultry Aerobic	Undercooked meat and poultry dishes, Unpasteurized milk	*Food infection* Small numbers of living bacteria required 100/g food	24–72 hours	Abdominal pain 'Flulike fever Diarrhoea Death rare
Staphylococcus aureus	Skin, nose throat of man causes septic skin conditions in man and other warm blooded animals Aerobic	Almost any moist food, cooked meats, ham, milk and milk dishes, custards, trifles	*Toxin poisoning* Enterotoxin formed in food before consumption	1–6 hours	Nausea Salivation Abdominal pain Violent vomiting Diarrhoea Sub-normal temperature Death rare
Bacillus cereus	Soil, vegetables, milk and cereals Aerobic: forms spores	Rice dishes, sauces Soups containing cereals	*Toxin poisoning* Toxin formed in food before consumption	1–8 hours	Abdominal pain Vomiting Diarrhoea Quick recovery
Clostridium welchii (perfringens)	Soil, sea mud, intestines of herbivores and man Anaerobic: forms spores	Soil covered vegetables meat and fish Improperly cooked dishes, stews and stocks	*Food infection* Large numbers of bacteria required 1 million/g food	8–22 hours	Abdominal pain Diarrhoea No vomiting Quick recovery Death rare
Clostridium botulinum	Soil sea mud Anaerobic: form spores	Semi-preserved meat dishes, sausages, pâté, pastes Home preserved low acid vegetables Home-cured hams	*Toxin poisoning* Neuro toxin formed in food before consumption	12–36 hours	Dizziness Double vision Change in voice Lassitude Paralysis Death 70%

In the United Kingdom recent cases have resulted from the consumption of tins of canned salmon which were either inadequately processed or the product was recontaminated after cooking by cooling water drawn into the can through a faulty seam. Commercially produced canned low acid foods, vegetables, meat and fish, may be regarded as safe because of the care taken in processing. Cans are heated in pressure cookers so that the centre of each can reaches 120°C for 2½ minutes. This will destroy botulinum spores unless present in abnormally large numbers. The acidity of fruits with pH less than 4.5 prevents the germination and growth of botulinum spores thus these items may be bottled or canned at the lower temperature of 100°C in safety. The use of additives in meat products can eliminate the risk of botulism. In the *Wiltshire cure method* for bacon the presence of nitrite inhibits the organism, nitrites may also be added to commercially produced pâtés. In sausages the use of sodium metabisulphite inhibits the growth of a range of food spoilage and food poisoning organisms including Clostridium botulinum.

CONTROL OF FOOD POISONING: FOOD HYGIENE

A number of factors are involved in an outbreak of food poisoning.

1 The food must be contaminated with pathogenic bacteria. Muscle flesh, milk and egg may become infected with Salmonellae in the living animal; vegetables will be infected with spore forming organisms from the soil in which they are grown.
2 Food handlers who are carriers of Salmonellae or Staphylococci can also spread their infection to food.
3 Careless handling of raw foods can spread contamination to items free of bacteria.
4 The initial contamination will be small but given time, warmth and nutrients the numbers of bacteria will increase to make the food poisonous to eat.
5 Preparation and cooking of food may aid the growth of some organisms, in releasing nutrients from tissues or in driving off oxygen and so create the necessary anaerobic conditions for some spore forming bacteria, heat at fairly high temperatures will activate germination of spores.

In two thirds of all cases of food poisoning investigated in recent years, the food had been prepared well in advance, that is 12 hours or more, before the time it was served. To this was added inadequate cooling and poor storage conditions and lack of refrigeration.

It is important that food poisoning is seen as a scourge which can be avoided by good kitchen management and the whole-hearted application of basic principles of food hygiene. The chief causes of food poisoning are given in the chart together with appropriate preventative action required to prepare food good to eat.

The overall conditions of kitchens, food stores, preparation and service areas must be taken into account. These must be of sound construction, clean, tidy, free of insect and rodent infestations and offer no hazard to food in the course of production. Thorough cleaning and disinfecting is required of equipment and surfaces with which food is likely to come into contact. An understanding of how micro-organisms may be destroyed is therefore required.

Table 6.2 *Top ten causes of food poisoning*

Causes	Action
Food used when contaminated with pathogenic bacteria	All raw food is subject to contamination Food poisoning organisms are ubiquitous Handle raw foods with care
Cross contamination of prepared food from raw foods	Separation of raw foods in storage and handling from foods ready for service Hygiene barrier for staff and equipment
Contamination of food by infected food handlers	Supervision and training of kitchen staff, hand washing routine, health screening
Food prepared too far in advance of service	Review work schedules 2 hour limit on foods in danger zone (10–63°C). Hot food to be held above 63°C Cold food to be held below 10°C
Cold food held at room temperature	Cold foods should be held at 5°C where possible. If temperature rises above 10°C for more than 2 hours foods should be discarded
Hot food not kept hot enough	Hot storage equipment must keep food above 63°C If temperature of hot food falls below 63°C for more than two hours it should be discarded
Frozen meat and poultry inadequately defrosted before use	Meat and poultry should be defrosted before cooking Joint size should be kept small about 4 lb (2 kilos)
Meat and poultry dishes undercooked	Meat and poultry should be cooked to an internal temperature of 65–70°C
Cooked food cooled too slowly	Cooked food to be held cold, should be cooled quickly to below 10°C in 2 hours
Inadequate reheating of prepared dishes	Temperature of reheated food should be taken to 70°C throughout the dish

THE DESTRUCTION OF MICRO-ORGANISMS

Methods for the destruction of micro-organisms can be divided into two groups:

Physical methods, such as heat and ionising radiations.
Chemical methods, the use of chemical agents as disinfectants.

Physical methods

1   Heat

The use of heat is the best method available. It is generally more reliable, less hazardous and cheaper than all other methods. It should always be considered first but its use may not be practicable in every case because of size or nature of a particular article. Heat at very high temperatures say 750+°C is instantly destructive of all life but such temperatures are destructive of most materials used in the kitchen except those made of metals such as iron and copper. Thus much lower temperatures must be used. However, there is an inverse relationship between time and temperature in the destruction of micro-organisms, as there is in most reactions involved in cooking food, thus if temperature is reduced then more time will be required to guarantee complete destruction of all micro-organisms and so ensure sterility. In catering practice sterility is very rarely required and we seek to reduce the numbers of micro-organisms present to very low figures, from millions to a few hundred. Thus temperatures of 100 and 120°C achieved by boiling and pressure cooking can be used effectively given the necessary time. The effects of heat will be affected by conditions in which heat is applied. The effectiveness of heat is increased by the presence of water. Everyone who has worked in a kitchen will know that a more serious burn will result from a short contact with steam from a boiling kettle at 100°C than hot air from an hot oven at twice that temperature, at 200°C. The presence of acids and alkalies promotes the effects of heat but organic substances especially proteins have a protecting action. Micro-organisms are not equally susceptable to the effects of heat, parasitic pathogens like Salmonellae are more readily killed by heat than non-pathogens. The spores of moulds and bacteria are more resistant to the effects of heat than vegetative cells, however mould spores are less resistant than bacterial spores. Very high numbers of cells or spores will increase the time required to ensure complete destruction.

2   Radiations

Electromagnetic radiations are destructive of micro-organisms as they are of other living things, the shorter wavelengths generally being the most effective. Thus the very short ionising gamma-rays and X-rays are the most destructive. Their effects are greater upon Gram vegative bacteria than Gram positive species. As might be expected, ionising radiations are less effective on spores than vegetative cells, bacterial spores being more resistant than mould spores. Use is being made of ionising radiations in the food industry in the destruction of mould spores in soft fruits and wider use is likely if acceptable to the public. Ultra-violet light is similarly effective and may be used to sterilize equipment and products in the pharmaceutical and medical field. The cleansing effects of sunlight are due to the presence of ultra-violet light. Longer wavelengths are not particularly destructive of micro-organisms in themselves but, as infra-red radiations and microwaves have heating effects, micro-organisms will be destroyed by the heat these radiations produce in water and foodstuffs.

Of all these ranges of electromagnetic radiations only microwaves are of immediate use to the caterer.

Table 6.3   *Food irradiation: recommended doses*

Purpose for treatment	Dose range (kGy)
Inhibit sprouting	0.05 – 0.15
Delay ripening of fruit	0.2 – 0.5
Insect disinfestation	0.2 – 1.0
Destroy parasites	0.5 – 6.0
Reduce bacterial load	0.5 – 5.0
Destruction of non-sporing pathogens	3.0 – 10.0
Sterilization	30.0 – 50.0

Chemical agents

Chemical agents, disinfectants and sanitizers, are widely used in the catering industry, thus it is important to understand the general principles involved in the choice and effectiveness of such agents. Chemical agents work through chemical action usually involving chemical union with a protein cell component such as a part of a membrane or enzyme molecule. The factors which affect chemical actions will also affect the action of disinfectants and sanitizers.

Factors affecting action of chemical agents

Temperature

Most chemical reactions are speeded up by heat. This is true of disinfectants and sanitizers. As a general rule 10°C rise in temperature will double the speed of action. Chemical agents should always be used in solution in hot water, as hot as is practicable keeping in mind the nature of the material under treatment and whether the agent is being applied by hand or by mechanical means. When cleaning by hand, temperatures of about 45°C are recommended, much higher temperatures of 80 to 90°C can be used in machine washing and cleaning. It is wasteful to use chemical agents in cold water.

Time
No agent works instantaneously, molecules have to collide with the cell of the target organism, diffuse through the cell wall, membrane and cytoplasm. This all takes time; just how much will depend on other factors of temperature and concentration. However, as a general rule, 10 to 15 minutes are required at 45°C and two or three at 80°C.

Concentration
For every chemical agent there will be an optimum concentration, a strength at which it works most effectively. This information will be available from reputable suppliers. It is wasteful to use more and dangerous to use less. Some bacteria can develop resistance to a disinfectant if they are subjected to sub-lethal doses of it. The misuse of a disinfectant can lead to a loss of potency against organisms which have grown accustomed to its presence. When using a chemical agent it must be diluted in accordance with the supplier's instructions; that means mixing a measured quantity of agent with a measured quantity of water. The habit of pouring neat disinfectant down the lavatory is profitable for the manufacturer but a waste for the caterer. Solutions of chemical agents deteriorate on standing so that solutions should be made up as required.

Inactivation
Chemical agents work by chemical action, by bonding with organic substances in the target cell. However, molecules of the agent will bond with other organic materials as well, such as food debris or even sink cloths and scouring pads. Also as the agent is absorbed by the cells of micro-organisms, should large numbers of cells be present, the strength of solution will fall below its effective strength. A solution which will kill a thousand bacteria may not destroy several millions. The solution of agent in water must make contact with cells and this may be prevented by a coating of fat or oil round the micro-organisms. The answer is clear, for success, when using a chemical agent a pre-wash is essential to remove heavy soil, ie obvious grime and dirt and food debris, and to reduce the bacterial load. Combined detergent sterilizers are available they are costly to buy and to use and may be ineffective in the presence of heavy deposits of grease and other food scraps.

Classification of chemical agents
There are many chemical agents and disinfectants on the market. In order to make an informed choice it is best to put compounds into groups and consider the advantages and disadvantages of each group.

Groups of chemical agents
    Phenols
    Chlorine and other halogens
    Quaternary ammonium compounds

Phenols

Phenol itself, distilled from coal tar was the first disinfectant and antiseptic used on a scientific basis although wine, vinegar and herb extracts had been used as salves in the treatment of wounds on an empirical basis for two thousand years or more. Phenol and closely related compounds are available as thick brown liquid concentrates such as carbolic and lysol. They are corrosive to the skin and have very strong pine-like odours. They are effective against a wide range of micro-organisms, bacteria, moulds and viruses but, because of their strong smell, they cannot be used anywhere near food which would quickly become tainted with the odour. They can be used outside in the cleansing of yards, refuse containers and drains. Less corrosive chlorinated phenolic compounds are available which give a milky solution in water, these white phenols have a marked sweet aromatic odour which again renders them useless in the kitchen but they are inexpensive and effective agents for use in accommodation areas, toilets and bathrooms.

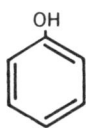
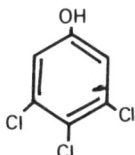
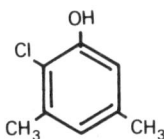
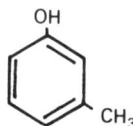

phenol   trichlorophenol   chloroxylenol   m-cresol

6.1   *Molecular structure of phenolic compounds*

Chlorine and other halogens

The halogens are a group of non-metalic elements which include the green pungent toxic gas *chlorine, bromine* – a poisonous pungent volatile blood-red liquid and *iodine* – a dark purple crystaline solid. They are sparingly soluble in water but soluble in dulute solutions of alkalies such as sodium hydroxide. These solutions are effective disinfectants. *Chlorine based* solutions are widely used, they are effective against a wide range of micro-organisms, bacteria, moulds and viruses. They have a seaweed like odour which can cling to equipment so adequate rinsing after use is required. They have good bleaching properties which are used in laundry work and they will remove colour from fabrics and furnishings. In prolonged use they will slowly remove the colour from ceramic sanitary fittings and strong solutions will bleach coloured plastic ware. Strong solutions will also etch metals and stain stainless steel. Because solutions are strongly alkaline they will also attack cellulose materials, cotton and wood. These problems can be overcome to some extent by the use of chlorinated trisodium phosphate, a stable crystalline material readily soluble in water. A similar material can be obtained but with the addition of sodium hypobromite so that a solution of both *chlorine* and *bromine* are obtained.

The properties of the organic halogen compounds are milder. Chloramines which slowly give off chlorine in solution are useful in the cleansing of delicate materials such as savoy bags where overnight soaking is recommended in a solution of the agent. Iodophores release *iodine* which is about ten times more effective than chlorine but it can produce brown stains if not used with care. Iodophores work best in acidic conditions, they are particularly useful in the cleansing of beer lines and pumps in bar and cellar work.

Quaternary Ammonium Compounds (QUATS)

These compounds are organic derivatives of salts of ammonia, which were used in the past as a cleansing agent. The principal base is ammonium chloride NH_4Cl, and in quaternary ammonium compounds the four hydrogen atoms are each replaced by organic groups either long or short chains or ring forms. Ammonium bromide may be used instead of the chloride. It is thus possible to make a wide range of quats. These compounds are colourless, odourless, tasteless solids, readily soluble in water with mild detergent properties. They are active against vegetative cells of fungi and both Gram positive and Gram negative bacteria, although Gram negative rods are less susceptible than Gram positive cocci. Because of these properties quats are useful sterilants throughout the kitchen but they are particularly recommended for glassware such as wine glasses where freedom from odour is essential. Chlorine based compounds may leave a seaweed like odour in drinking vessels objectionable to the connoisseur, whereas quats rinse off well leaving no taint. Quaternary ammonium compounds are cationic detergents and their action is affected by the presence of anionic compounds such as soap.

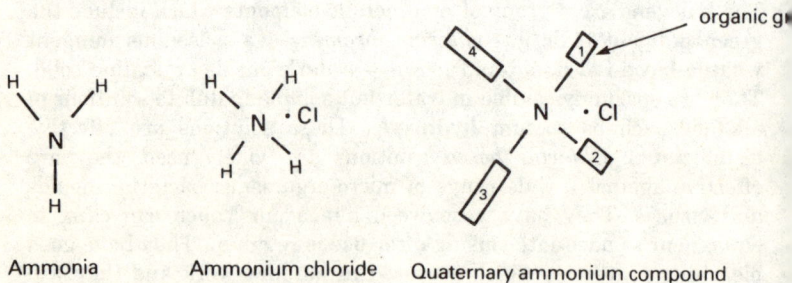

Ammonia   Ammonium chloride   Quaternary ammonium compound

6.2   *Molecular structure of quaternary ammonium compounds*

OTHER DISEASES SPREAD BY FOOD AND WATER:

Mycotoxicoses

It has already been stated that some moulds produce toxins when growing on food. Should this mouldy food be eaten then disease may result. As long ago as 1597 it was shown that the disease St Anthony's Fire or *ergotism* was caused by eating bread made from mouldy rye, yet outbreaks of the disease continued into the twentieth century. In 1961 it was shown that a disease amongst turkeys, Turkey X disease, was caused by feed containing mouldy peanuts. In the last 25 years evidence has mounted that many species of moulds produce a variety of substances toxic to man and other animals, both the genera Penicillium and Aspergillus contain toxinogenic species. Mycotoxins may cause acute and fatal poisoning such as Islanditoxon from Penicillium islandicum but more often they are slow and insidious in their effects and as the toxins accumulate in the tissues they may affect liver, kidneys, bone marrow or nervous system. Mycotoxins will diffuse through a food material on which a mould is growing thus merely to cut out or scrape off mould growth is of no avail. Mycotoxins produced will be in the depth of the product which is best thrown away. To have or to use food known to be mouldy in the kitchen would be an offence which could lead to prosecution. The position of mould ripened foods, such as blue cheeses, is anomalous. Not all moulds are toxinogenic. Penicillium roqueforti used to make Roquefort, Stilton and Gorgonzola is thought to be harmless. However, reports that over-indulgence in such cheeses leads to nightmares suggest the presence of neurotoxins.

Legionnaires' disease

Legionnaires' disease is not strictly a food borne intestinal disease, it is a type of pneumonia which may be accompanied by nausea, vomiting and diarrhoea. However, it is a disease of interest to the caterer because it is associated with hotels, hospitals and institutions. The organism involved is Legionella pneumophila, found as Gram negative motile rods, aerobic and mesophilic, widely distributed in the soil and water. Respiratory diseases, such as pneumonias, are droplet infections where the organism is carried in minute water droplets suspended as an aerosol in the atmosphere. Such aerosols would normally be formed by the respiratory movements, coughing and sneezing, of infected suffers. In the case of Legionnaires' disease the bacteria laden aerosols may be formed by contaminated water from sprays in humidifiers, air-conditioning systems, shower-baths and even from ordinary taps. As a general guide, air-conditioning plant and humidifiers should be inspected, drained, cleaned and disinfected with a solution of chlorine at 50 parts per million of water every six months. Similarly shower-bath spray units should be dismantled and their parts sterilized regularly. It is important that hot water is kept hot and cold water kept cold, thus cold water tanks should be protected from heat in summer months to hold water at about 15°C. It is known that the organism can survive in water pipes for more than a year.

Intestinal parasites

Intestinal parasitic worms acquired by the consumption of under-cooked meat have been known for several thousand years. Measures for their control may be reflected in the ancient Hebrew food laws of

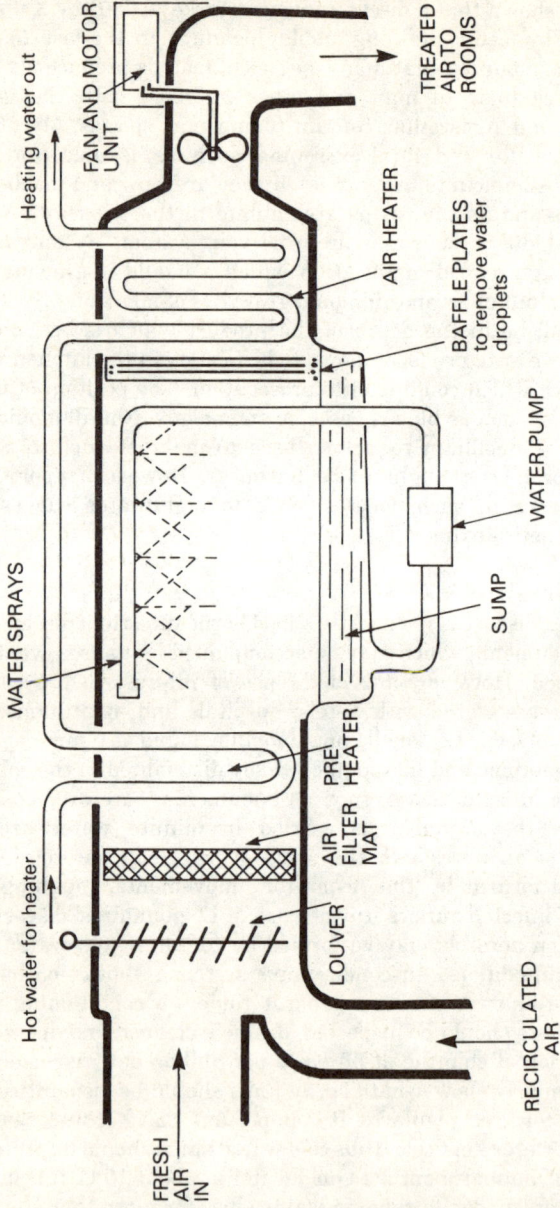

6.3   Diagram of air-conditioning plant

the Torah. Although a relatively minor problem, the evidence is that infection is increasing. The tapeworm is the chief parasite of which there are two species:

> Taenia saginata – the beef tapeworm
> Taenia solium – the pork tapeworm.

The life cycle of the tape worm is in two parts: the adult stage in man and a larva stage in the meat animal. To begin the cycle, the meat animal must eat fodder contaminated with ova shed in the faeces of a human infected with an adult worm. When the ova reach the duodenum of the animal, each embryo hatches from its egg-case and the larvae make their way through the intestinal wall into the blood stream and into the muscle tissue. In the muscle tissue larvae form cysts, 3 or 4 mm in diameter, containing the invaginated head of the adult worm. These cysts or cysticerci may be seen in 'measly' meat on inspection. The cycle is completed when infected meat, which has not received adequately cooking, is subsequently eaten. As the meat tissues are digested in the small intestine of the human host, the cyst evaginates and the head of the adult worm, the scolex, attaches itself to the intestinal wall by means of hooks and suckers and grows into a full sized adult worm which can reach a length of more than a metre. The body of the worm is made up of a ribbon of individual segments or proglottids, each complete with male and female reproductive systems, thus fertilisation of ova is ensured. Proglottids are produced continuously from the 'neck' of the scolex and the older segments containing fully developed fertile ova, the gravid proglottids, at the far end, break away and pass out of the host's body and are voided with the faeces. Insanitary disposal of sewage or use of faeces as fertilizer may result in the contamination of pastures with tapeworm ova and the cycle is primed to begin over again. In the United Kingdom it is believed birds such as sea gulls, pecking over sewage sludge, pick up gravid proglottids then shed many hundreds of ova, which pass, unaffected, through the birds' intestines and in their droppings onto grazing land. Cooking meat to an internal temperature of 60°C will destroy the cysticerci. Holding meat frozen at −15°C for three weeks is also satisfactory treatment.

Another still rarer parasite is a small nematode worm, Trichinella spirallis. This parasite completes its life cycle in one host. The host which can be man, pig or other mammals such as whale or polar bear, acquires the infection by eating larvae in cysts in uncooked infected meat. Larvae, released by the digestion of meat tissues and the cyst walls, invade the intestinal wall of the host. During the course of a week or so the larvae mature to adult male or female worms. In some three weeks the fertilized female will release living larvae into the blood stream. The larvae travel to the muscles and there form further larval cysts in the host's tissues. The invasion of the intestinal mucosa will cause nausea, vomiting and diarrhoea one or two days after eating infected meat. The migration and cyst formation may cause fever and muscular pain. Adequate cooking or deep freeze storage results in the destruction of Trichinella.

ADULT WORM in man as host

gravid proglottids and fertilized ova shed in faeces

seagulls pick up eggs in sewage sludge

eggs in gulls droppings

fertilized ova assimilated during grazing

LARVAE hatch in intestines of bovine

LARVAE migrate form cysts in muscle (cysticercus)

undercooked meat eaten by man evagination of cysticercus

hooks

suckers

head (scolex) of adult worm

longitudinal canal

uterus

testis

genital pore

ovary

vitelline gland

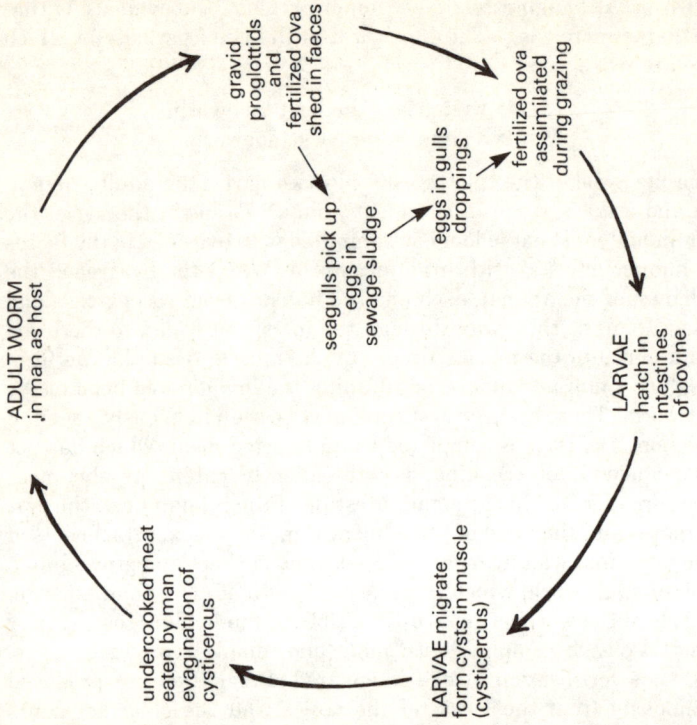

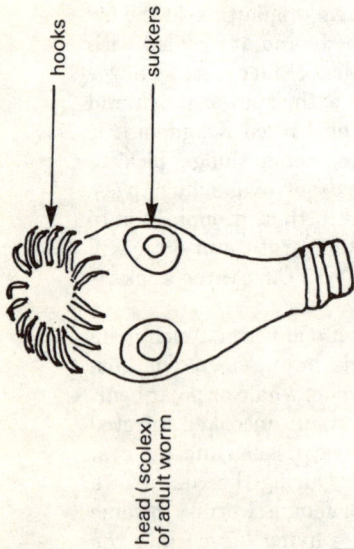

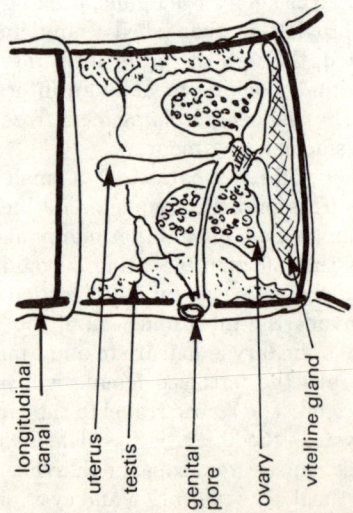

mature segment (proglottid)

KITCHEN PESTS

Aesthetics reinforced by food hygiene regulations require that food is prepared in kitchens free from infestation with insect and rodent pests.

INSECT PESTS

There are a number of insects pests and they may be divided into two groups according to methods of locomotion that is:

Crawling insects – example: cockroaches.
Flying insects – example: the house fly.

Cockroaches

These insect pests are regarded as a hazard to health because they crawl between drains, sewers, refuse bins and kitchens in search of food and shelter. They foul food stuffs, as they crawl over them, with their excreta, regurgitated stomach contents and unpleasant smelling pheromones. Cockroaches can harbour a wide range of micro-organisms including food poisoning species Salmonella and Staphylococcus. Salmonellae can survive in cockroach faecal pellets for more than three months especially in dry weather. There is no doubt that these insects are well able to spread disease. There are three species of interest:

1 *Blattella germanica*   The steam fly or German cockroach: light brown in colour, about 15 mm in length.
2 *Blatta orientalis*   The black beetle or oriental cockroach: dark brown to black in colour, about 30 mm in length.
3 *Periplaneta americana*   The American cockroach: chestnut brown in colour, about 40 mm in length.

In principal the life cycle for each species is much the same. The fertilized female lays her eggs in a purse about 5 mm × 2 mm. Each purse will contain 16 eggs in two rows of eight; 40 in the case of Blattella germanica. She will carry the purse for some days and then push it into a warm crevice, cementing it firmly in place. The eggs will hatch in two to three months according to the local temperature. The young emerging nymphs are miniature versions of the adult. They grow to full size and sexual maturity in two to ten months depending on local temperature. As they grow, the nymphs must shed their tough exoskeleton which cannot grow with them. This moulting or *ecdysis* occurs six times in the growth period and the shedded cuticle can contaminate food and provide embarrassing evidence of cockroach infestation.

Blattella germanica likes a hot steamy habitat and is to be found behind and under sinks, ovens and in the lagging on hot water tanks, particularly if dampened by leaks. It is an active climber and can scuttle up tiled or painted walls.

Blatta orientalis tends to live 'below stairs' in boiler rooms and associated ducts and pipes, in subterranian kitchens and stores.

Periplanita americana is a more recent arrival than the oriental cockroach which has been in Britain since the fifteenth century. However, it is now to be found in warm buildings such as breweries, bakeries and warehouses, particularly in ports and coastal towns.

Signs of infestation

Cockroaches are nocturnal, thus it is possible for a kitchen to be infested without insects being seen by kitchen staff during the working day. Signs to look out for are droppings, regurgitated food, moulted skeletal fragments and cockroach odour. Cockroach traps – small cardboard cartons containing bait or a sticky surface to hold wandering insect pests for inspection in the morning – are very useful in monitoring infestation control.

Infestation control

1 Ensure that the kitchen, food production and storage areas are kept clean and that food debris does not build up on working surfaces and under tables, ovens, refrigerators and other pieces of equipment. Clear out all refuse at the end of each shift.
2 Keep building structure in good repair. See that all cracks, nooks and crannies are filled in. Ideally, equipment should be mounted away from the walls to leave a gap of 500 mm for cleaning purposes.
3 To remedy an infestation, harbourages may be treated with a wettable powder containing either organophosphorus or carbamate insecticide. Insecticidal laquers containing dieldrin may be applied in strips about 100 mm wide around infected areas. It will be necessary to repeat treatment twice at intervals of two months to ensure success.

Very great care must be taken to see food stuffs are not contaminated with insecticidal powders, sprays and lacquers. There are well-known pest control organisations from whom reliable professional help may be obtained.

The house fly

There are more than a hundred species of two-winged flies in the United Kingdom. However, in the kitchen, the most troublesome is the housefly, *Musca domestica*. It is a serious menace to health as it is an unselective omniverous feeder, wandering from filth to food stuffs carrying dirt and disease wherever it does. The body of the adult fly is covered with hairs. It has six hairy legs ending in pairs of claws. Under each claw is a glandular pad which secretes a sticky fluid. It is by means of these sticky glandular pads that the fly can hang upside down on ceilings, light-fittings and similar surfaces. The fly's hairy appendages and sticky pads pick up particles of filth and become heavily contaminated with micro-organisms, bacteria, moulds and viruses. Investigations have shown flies carry a load of between two and three million bacteria each. The adult fly has no biting mouth parts but the mouth parts form a tube-like proboscis ending in twin pads. Secretions containing digestive enzymes are regurgitated from the stomach onto food and spread over its surface by a squeegee-like action of the pads at the tip of the proboscis and then sucked back into the stomach. Thus, when feeding in say a sugar bowl, a fly will regurgitate its stomach contents perhaps acquired from an earlier visit to a dung heap, the possibilities for cross-contamination are obvious.

Life cycle of house fly
The female fly will lay eggs from about the third day after reaching sexual maturity. Eggs are laid in batches of about 50. A female may lay five or six batches in warm moist material, preferably horse manure, cow dung or midden heaps but kitchen refuse and spent tea leaves will also serve. Eggs take about 12 hours to hatch when conditions are most favourable, the larvae or maggots are about 1 mm long and, feeding voraciously, grow to 8 mm in four or five days. The larvae then migrate to secluded cracks and crannies to pupate. The pupa is a brown barrel shaped chitinous structure about 5 mm long. Within the puparium metamorphosis takes place, the larval structures are digested and from the jelly-like mass the adult fly forms. Metamorphosis is completed in four days in high summer or the insect may rest through the winter and emerge in the spring to breed the new season's pests.

Infestation control

1   Refuse should be held in tightly lidded bins which should be emptied daily if possible. Bins should be sited as far away as is possible and reasonable from kitchens. Bins should be washed out regularly.
2   Open windows of kitchens, food production and storage areas and air vents should be screened with wire gauze (BSS 24 mesh)
3   Where outer doors must be left open for the movement of goods, these should be screened by an air-curtain from a strong fan mounted over the doorway. Flies are unable to fly against fast moving air currents.
4   Flies swarming in kitchens, serveries and like areas may be destroyed by use of aerosol sprays of low toxicity such as pyrethrum synergised with piperonyl butoxide. Food items must be protected from contamination. Dead and dying insects must be swept up. Surfaces with which food may come into contact must be washed over with hot water and detergent.
5   Apparatus consisting of an electrified grid and a violet light attachment, attractive to flying insects, has been found useful in some cases. The insects fly towards the light into the grid and are

electrocuted, their bodies falling into a tray suspended below. Such units must be sited with some thought if they are to be successful and placed where flies are seen to gather. Such units must be cleaned weekly at the very least.

RODENT PESTS   Figure 6.5

The family of rodents includes mammals which gnaw (Latin: *rodere* = to gnaw), such as rats and mice, as well as beavers, rabbits and squirrels. Infestation by rats and mice is a serious matter not only for the economic loss of raw materials eaten or contaminated with hairs, faeces and urine but also because rodent pests spread a number of diseases which include typhoid fevers, Weil's disease and bubonic plague. They may also damage property and furnishings as a result of their burrowings and gnawings.

There are two important species of rat.

1   *Rattus rattus*: the black rat or ship rat
2   *Rattus norvegicus*: the brown rat or sewer rat.

The black rat: Rattus rattus

The black rat is the smaller of the two species, some 200 g in weight with body length about 200 mm, and a longer tail about 250 mm. It has large translucent ears and a pointed snout. In colour it may be black, brown or grey. It is found in the holds of ships, dockside warehouses, food factories, hotels, restaurants and institutions in ports and large towns. It is an active climber so generally nests in lofts and attics. Adults will consume about 12 g per day.

The brown rat: Rattus norvegicus

The brown rat is the larger of the two species, some 350 g in weight with body length about 250 mm and a thick short tail about 200 mm. It has small hairy ears and a snub-nosed snout. In colour it may be brown, grey-brown or brown-black. It is found throughout the countryside. It is a burrower and a good swimmer so will be found not only in burrows in river and canal banks and in sewers but also in railway embankments, rubbish tips and about farms and factories wherever food is available. The adult rat will consume about 25 g per day.

The house mouse: Mus musculus

The house mouse is not as serious a pest as are rats, however they will contaminate food with their faeces and urine and sight or smell of them will lead inevitably to loss of customers' goodwill. In appearance the mouse is a small replica of the black rat with pointed snout, large translucent ears and long slender tail. However it weighs only some 20 g with body length 80 mm and tail 10 mm. In colour it may be light grey or light brown. It is both a burrower and a climber and somewhat erratic in its habits, so may be found almost anywhere where there is food and shelter. The adult mouse will consume about 3 g per day.

Rattus rattus

Rattus norvegicus

Mus musculus

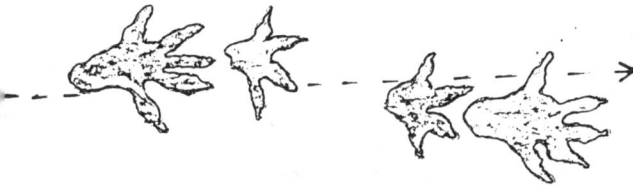

Tracks of Rattus norvegicus

6.5 Rodent pests

Signs of infestation

Rodents are wary of man and are generally nocturnal in habits. It is possible for an hotel, restaurant or institution to harbour a family of rats and yet no rats are to be seen. Only when the colony is a large one of 20 or more is it likely that rats will be seen in daylight hours. The sight of droppings are usually the first signs, then foot prints in dusty surfaces. Flour or talc can be put down to detect the presence of rats in this way. Being rodents they may be expected to gnaw at the woodwork of doorways, packing cases and cupboards to get at food materials. A well established colony will have definite trackways marked by greasy smears with embedded rodent hairs left as they slink over pipes, cables and girders and round corners and doors.

Table 6.4   *Signs and size of rodent infestation*

Signs	Less than 20	Size of colony Between 20–50	More than 50
Live rats	May not be seen in day	Single rats to be seen in day	Numerous rats seen in day
Pellets	Small numbers of pellets of same size in 4 or 5 places	More than 30 pellets 2 sizes in 10 to 15 places	Many pellets 3 sizes in more than 20 places
Gnawings	Little sign 1 or 2 places only	10 to 20 areas in woodwork of food stores	Many gnawed holes in doors and floors
Tracks	May not be seen only by dusting techniques	Tracks found in dusty areas on cupboard-tops and the like	Many clear track marks
Runs	None	Rarely found	Definite grease marked runs

Control of rodent infestation

The rat population of the United Kingdom is estimated to be over 50 million, that is there is one of them for every one of us. Their complete eradication is well-nigh impossible, thus the caterer must adopt a siege-like attitude and establish in his own premises a 'no-go-area' for rats. This means constant vigilance and the type of buildings which offer no easy access to shelter or supplies of food or water.

Control of access

Doors and windows   Doors and windows should fit well within their frames without gaps, especially at the bottom of doors which are best sealed with a nylon bristle strip. Metal kick-plates and angle-iron should be used to protect the foot of doors and door-jambs from attack by gnawing. Ground floor window openings should be protected by wire-mesh screens.

Pipe-work   The outer brickwork must be kept in good condition. Holes made for pipes, ducts and cables must be sealed with cement, often neglected by builders. Traps on sewer pipes must be checked for effective operation and manhole covers and drain grids should be sound and snug-fitting. Sewer vent pipes should be closed with a wire-mesh balloon cap.

Storage of food materials

All food stuffs and edible refuse should be stored in rodent-proof containers. This is particularly important if stores are outside the main building. Rubbish should not be left to accumulate but disposed of often and regularly.

Eradication of infestation

If an infestation is discovered it is best to seek assistance from professional pest control agents, however DIY rodenticides are available. The most widely used rodenticide in recent years has been *warfarin*, an anticoagulant which causes death by promoting internal haemorrhage. However, both rats and mice in some areas show resistance to this agent so that alternative materials may best be used. *Alphachloralose* is recommended against mice and *brodifacoum* or *bromadiolone* against rats. Rodents may be suspicious of materials put down for them and show 'bait shyness', thus leaving the poison untouched. The poison should be mixed with some foodstuff normally available in the kitchen which they may have been taking, few rodents can resist succulent sultanas for example. The poisoned food should be put into bait trays to prevent it being scattered during feeding. The trays should then be placed where the runs are thought to be and left undisturbed for four or five days then replenished if necessary. A problem with the use of poisons is that rodents will return to their harbourages to die, if these are in the building under floors, behind skirting boards and in wall cavities then the subsequent purtifaction of carcases could result in unpleasant odours. All rodenticides must be stored, handled and mixed outside food storage and production areas. Bait trays must be clearly marked '**POISON DO NOT TOUCH**' and so placed that there is no risk of contaminating food for human consumption. Staff working in baited areas should be told that rodenticides are being used. If poison is mixed with normal food ingredients, only that required for immediate use should be prepared.

EXERCISES

Food poisoning results from bad kitchen practices. What type of poisoning could arise from the following incidents? Name the organism likely to be involved, giving your reasons fully.

1 Rubbing the nose when making ham sandwiches.
2 Serving lightly cooked mushrooms which had been kept a week raw in a polythene bag.
3 Eating steak pie kept warm in a defective hot cupboard.
4 Spit-roasting frozen chicken not fully defrosted.
5 Making a mousse from ducks' eggs.
6 Serving peach condé at a buffet on a hot summer night.

7

The Structure of Food Materials

All foods are derived directly or indirectly from plants or animals. Many of these food items are eaten with little preparation so that their natural structure will be the major factor governing their eating qualities. An understanding of the structure of plant and animal tissues is thus important in the choice, preparation, cooking and service of natural food items, so that their qualities may be retained to be enjoyed by the consumer.

THE STRUCTURE OF VEGETABLE TISSUES

In every-day language we speak of fruit and vegetables as if the two were different classes of foods, whereas all materials taken from plants are vegetables. Fruits are special vegetables; they are reproductive structures normally bearing seeds when fully ripe. Runner beans are unquestionably fruits but rhubarb, enjoyed by some as a fruit, is a sour-tasting pigmented leaf stalk or petiole. The tissues found in fruit are similar to those found in other vegetable structures.

As a wall is made up of bricks so are plant and animal tissues made up of cells and, as the properties of a wall reflect the properties of the bricks and morter used in its construction, so the properties of a food item will reflect the properties of the cells and their cohesion within the tissues of that food item.

The plant cell

The plant cell is a box-like structure, sizes and shapes will vary between 100–500 μm, some cells almost spherical others long and thin like pencils.

The outer non-living cell wall, more or less rigid according to thickness, is composed of *cellulose*, one of the polysaccharide. Inside the cell wall is the living cell enclosed in a membrane largely phospholpid with some protein. Inside the membrane there is a jelly-like material similar to egg-white, about 12% protein and 88% water, called *cytoplasm*. Floating in the cytoplasm there may be minute capsules of lipoprotein called *plastids*. There are three types of plastids:

1 *Chloroplasts*: these capsules contain the green photosynthetic pigment chlorophyll, and are found in leafy tissues.

2 *Chromoplasts*: these capsules contain the water-insoluble orange carotene pigments. They may be found with chloroplasts in leafy tissues and in pigmented structures such as tomatoes and carrots.

3 *Leucoplasts*: these are colourless capsules and contain starch. They will be found in leaves during photosynthesis and also in very large numbers in storage organs, roots, tubers and seeds, such as turnips, potatoes and peas.

Together with such plastids each cell will contain a nucleus of genetic material, DNA enclosed in a membrane also floating in the cytoplasm. A major feature of each cell will be a fluid filled cavity, the *vacuole*, enclosed in a membrane, the *tonoplast*. The fluid, called *cell sap*, is a solution of sugars, salts, amino acids and water-soluble pigments. The amount and composition of cell sap will be reflected in the succulence of a fruit or vegetable, and the hydrostatic pressure of the fluid will control the turgidity of the cell and the firmness of the tissue. The cell sap is analogous to the air in a tyre. If air is lost the tyre becomes soft and flaccid, similarly if water is lost from a cell it too becomes soft and flaccid. The crispness of a lettuce leaf is due to the hydrostatic pressure of water in the cell sap, if that water is lost, crispness is lost and the leaf becomes soft and limp. Within a fruit or vegetable there will be different types of tissue present, each making a contribution to the

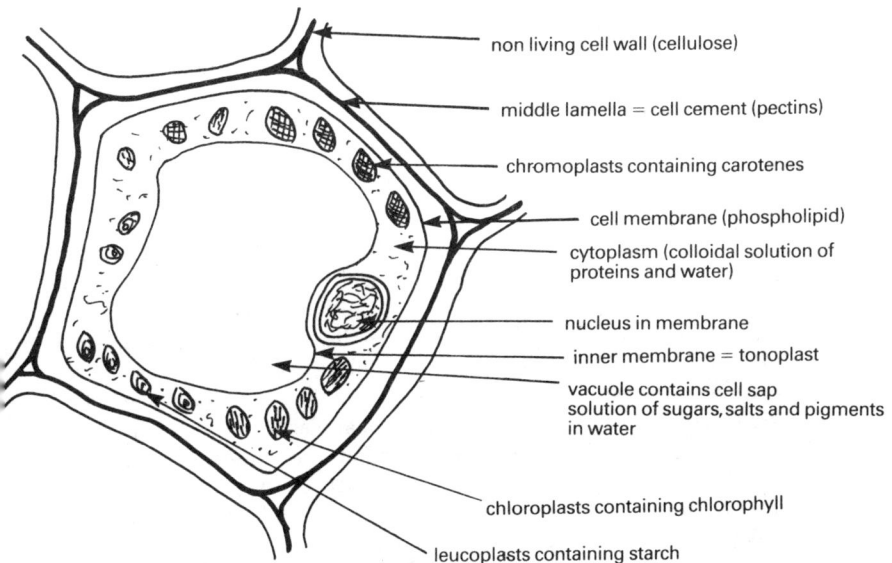

non living cell wall (cellulose)

middle lamella = cell cement (pectins)

chromoplasts containing carotenes

cell membrane (phospholipid)

cytoplasm (colloidal solution of proteins and water)

nucleus in membrane

inner membrane = tonoplast

vacuole contains cell sap solution of sugars, salts and pigments in water

chloroplasts containing chlorophyll

leucoplasts containing starch

7.1 *Structure of plant cell*

over-all eating quality of the item. The cells in each tissue will be modified in structure to fulfil a particular function of the tissue. There are six types of tissue:

1 *Epidermis*. This forms the protective outer skin.
2 *Parenchyma*. This forms the soft and juicy parts.
3 *Collenchyma*. This is found as columns of tough cells supporting the softer parenchyma.
4 *Schlerenchyma*. This is similar to collenchyma, tough supporting columns of cells associated with the softer vascular, water conducting tissue phloem.
5 *Phloem*. This, together with xylem, form the vascular or water conducting tissues. Phloem carries sap, water and sugars, from the leaves down the plant to structures below. Xylem carries water and salts absorbed by the roots up through the plant to the leaves, flowers and fruits above. Phloem is a soft tissue.
6 *Xylem*. This is made up of tough dry tube-like cells.

Epidermis

The cells of the skin of vegetable structures are regular brick-like units and fit together, with no spaces between them, to give a dense compact layer. The outermost walls will be thickened and coated with a waxy cuticle. As the vegetable ages so the walls become thicker and the waxy deposit impregnates the cell structure. Thus with mature items it may be necessary to remove the skin to make the item edible.

In young immature structures the skin is readily eaten and need not be removed. In catering practice the epidermis should be left intact unless it is too tough or unsightly to be eaten by the consumer. Removal of the epidermis will lead inevitably to the loss of nutrients and turgidity and thus some loss of quality.

Parenchyma

The cells of the fleshy tissue, parenchyma, are large, polyhedral or almost spherical in shape, with thin cell walls. The cytoplasm also forms a thin layer inside the cell and the bulk of the cell is occupied with a single large fluid-rich vacuole. As the cells of the tissue press up against each other, air-filled spaces will be left where adjacent cell walls do not meet. Such air spaces may make up 1% of dense potato tissue and 25% of the less dense tissue of apple. Cells are held together by a layer of pectin cell cement called the *middle lamella*. The integrity of this layer will control the texture of the whole tissue, so, for example, if the cement is dissolved on boiling, the tissue will collapse. The difference between varieties of potato is due in part to this effect. Floury varieties such as King Edward and Maris Piper give a soft floury texture on cooking because of a loss of adhesion between the cells of the parenchyma. Waxy potatoes such as the Pentland varieties give a firm to flinty texture on cooking and the middle lamella retains much of its adhesiveness.

Parenchyma may also serve as a storage organ, when the cells will contain large numbers of starch-filled leucoplasts. The cooking time of vegetables will be governed by this. If leucoplasts are few then cooking time will be short – five minutes or so but if leucoplasts are present in large numbers the cooking time will be longer – 15 or even 25 minutes. The longer cooking time is required to gelatinize the starch held in the leucoplasts.

Collenchyma

This tissue is found as rod-like columns of cells set into the matrix of parenchyma to give support to the softer tissue, the ridges which run up the outer face of a celery stick (*petiole*) are due to such columns of cells just under the epidermis. Cells of collenchyma are long and thin. In cross-section they tend to be roughly pentagonal fitting tightly together. Their cell walls are thickened particularly at the corners. This thickening of the cell wall during the life of the cell results in a steady change in the nature of the tissue so that when the tissue is young it is soft and readily eaten but with age the tissue becomes more and more resiliant and chewing becomes difficult so that an intractable mass of threads remains in the mouth after the softer tissues have broken away and been swallowed. This problem is seen in celery and rhubarb.

Sclerenchyma

This tissue has a supporting role like that of collenchyma. It is usually found as bundles of rod-like cells supporting the soft phloem of the vascular system. The cell walls of this tissue are uniformly thickened and as the tissue ages the cell walls become woody or lignified and as the cells die their fleshy contents are lost, resulting in the tissue becoming tough and dry. This defect can be found in the various types of bean where the fruit case is eaten. If harvesting is delayed the sclerenchyma associated with the vascular bundles running down the pods becomes lignified so that the vegetable becomes stringy and more and more fibrous so that when it is eaten sharp woody splinters pierce the tongue, cheeks and gums. The fibrousness of asparagus is caused in the same way.

Sometimes small clumps of sclerenchyma cells can be found in soft parenchyma. These clumps, called *schlereids*, are found in some varieties of pear and in quince, and are responsible for the gritty texture of these fruits. Sclereids may also be found in nuts and in the skins of pulses.

Phloem

This soft tissue has the important role in the life of the plant of conducting the products of photosynthesis from the leaves to the rest of the organism. However, the tissue plays little part in the eating quality of fruit and vegetables. The cells are generally long, thin and fleshy with thin cell walls, dense cytoplasm and small vacuoles. Some cells called *sieve cells* have their end walls at the top and bottom punctured by tiny holes. These cells form a long series of conducting tubes but they retain their living cell contents for the life of the tissue.

Xylem

This tissue is the second of the two types of vascular tissue. The cells are long thin pencil shaped cells, tightly packed together. As the tissue ages the walls forming the long sides of the cells become thickened and lignified, the end walls are broken down and the living cell components lost. The tissue then becomes a system of dead empty woody tubes for conducting water and mineral salts. The ageing of xylem may result in the loss of eating quality in root vegetables such as carrots and parsnips where, towards the end of the season, the vascular tissue becomes markedly woody and no amount of cooking can make the product edible.

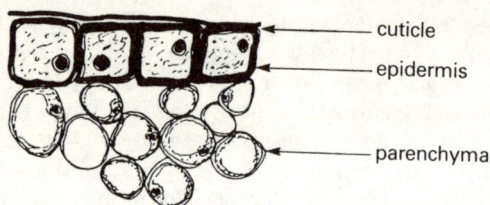

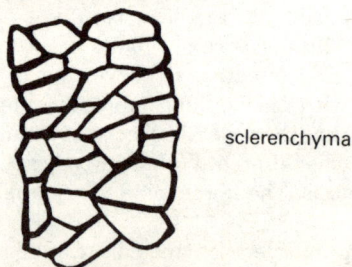

7.2   Structure of cells of plant tissues

THE STRUCTURE OF ANIMAL TISSUES

For many of us meat is the major food item. It is the item to which we give the most thought and care in purchasing. It is probably the most expensive but a source of lasting pleasure in eating for its unique range of textures, flavours and nutritional qualities. Meat is however a complex tissue derived from the muscles of domestic animals.

The animal cell   Figures 7.3 and 7.4

In its simplest form the animal cell may be seen as egg-shaped or perhaps lens-shaped about 100 μm in diameter as, for example, the lymphocyts in blood or the cells from the inside of the mouth. At first sight under the ordinary microscope they appear less complicated than plant cells and show three main features only. These are: a *cell membrane*, a large *central nucleus* enclosed in a membrane and a mass of granular *cytoplasm*. Vacuoles, if present, are very small. Under the electron microscope at magnifications or 100.000 times a far more complex structure is to be seen:

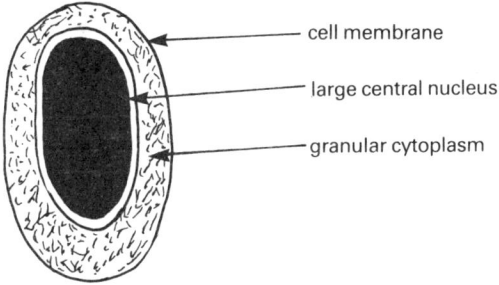

cell membrane

large central nucleus

granular cytoplasm

7.3   Lymphocyte: a simple animal cell

1   There is a maze of very fine tubes which run through every part of the cell radiating out from pores in the nuclear membrane. This network of canals provides the means of communication between the nucleus and other parts of the cell and chemical messengers are sent from the nucleus to activate processes in other organelles. This canal system is called the *endoplasmic reticulum.*
2   Attached to the endoplasmic reticulum in the outer parts of the cytoplasm are minute spherical bodies about 0.1–0.5 µm in diameter called *ribosomes*. These organelles are responsible for the synthesis of protein molecules required in the cell, enzymes for example.
3   Scattered through the cell are sausage-shaped bodies about 2–5 µm in length called *mitchondria*, these have the function of energy management and the synthesis of adenosine triphosphate, the energy rich material used in the cell to power all its activities.
4   Together with the mitochondria are somewhat smaller spherical bodies, the *lysosomes*, which contain digestive enzymes required in the metabolic processes of the cell.

The structure of meat
Meat is made up of two tissues: the *muscle cells* which have the essential property of contraction in the living animal and *sheets of connective tissue* which bind the cells together into a working unit and connect the muscle to the limb bones on which the muscle operates. In a piece of meat it is not possible to separate the two tissues which must be eaten together, thus the texture of a piece of meat depends upon the combined properties of the two tissues. Indeed the use to which a cut of meat can be put, the preparation

and cooking required, will largely be dictated by the amount and nature of the connective tissue present.

The structure of connective tissue   Figures 7.5 and 7.6

In its simplest form connective tissue is best seen in the form of a thin translucent structureless sheet of glycoprotein supported by a network of fine elastic threads which branch and join together in an irregular manner. This network is composed of the protein *elastin* and is reinforced with wave-like coils of stouter non-branching threads of the protein *collagen*. There may be a third group of fibres of reticulin similar to collagen. The protein threads are accompanied here and there by the cells which fabricate them, the *fibroblasts*. The chemical nature of the two sets of fibres is of great importance to the caterer and will determine how the meat is to be cooked. Collagen is degraded by heat at temperatures between 80–100°C and undergoes hydrolysis and converted to water soluble gelatin. Elastin is resistant to cooking. The amount of collagen present in a joint of meat will dictate the manner and length of cooking required. The amount of elastin will determine, together with the state of the muscle cells, the toughness of the final product.

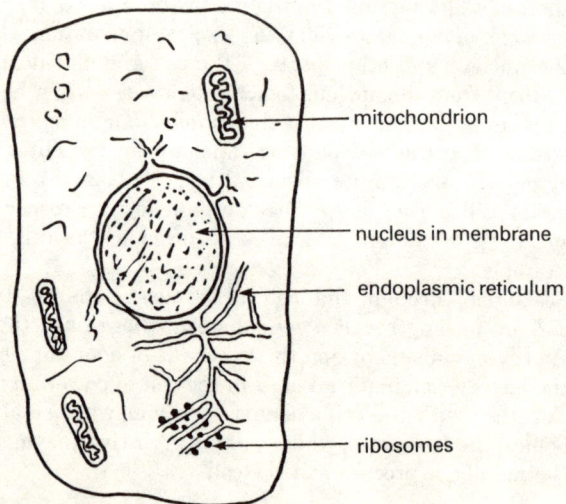

7.4   Detailed structure of animal cell

The spaces between the mesh of threads is filled with glycoprotein and embedded in this structureless matrix are numbers of tissue cells some of which have the function of fat storage. Thus according to the state of nutrition some of the connective tissue cells will be swollen with fat droplets to five or ten times their original size. The presence of the fat will give a dense white or creamy appearance to the connective tissue, termed *marbling*. The amount of fat in meat is a matter of personal taste. Fat is responsible for species flavour, that is the flavour of beef is due to the presence of beef fat and that of mutton to the presence of mutton fat. The presence of fat dilutes the concentration of tissue fibres and thus promotes tenderness. In the act of eating, as the teeth burst the fat-filled cells, the released oil lubricates the chewing process and flows over the surface of the gums to give a feeling of well being and delight. However, to most people excess fat is cloying.

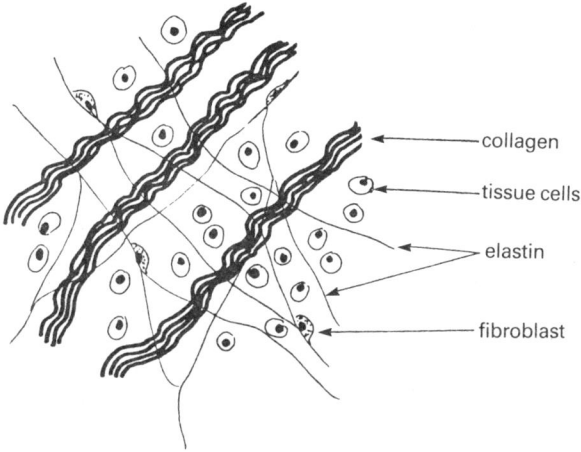

7.5   Structure of areolar connective tissue

The structure of muscle tissue
Muscle is not made up of cells as in other tissues but consists of bundles of long thin multinucleat fibres. These fibres will vary in length according to the length of the muscle from 5 mm to 500 mm and are about 10–100 μm in diameter. But this will vary from muscle to muscle, with age, exercise and levels of nutrition.

The muscle fibre   Figures 7.7 and 7.8
Surrounding the fibre is a membrane, the *sarcolemma*. Inside is a mass of a semi-fluid jelly-like material, the *sarcoplasm*, in which a number of nuclei are found. In the centre of the fibre is a stout bundle of fine threads, the *myofibrils*, each about 1 μm in diameter running the length of the fibre. The myofibril is divided along its length into individual units about 3 μm in length. These units are called *sarcomeres* and are the individual contracting units of the muscle. The dividing walls between adjacent sarcomeres are termed *Z-plates*. Attached either side of the Z-plate are bundles of protein threads, *actin* threads, about 2 μm in length. In the centre of each sarcomere is a bundle of somewhat stouter protein rods, *myosin* rods, about 1.6 μm in length. The sarcomere is able to contract because the actin threads can run into the spaces between the myosin rods. Each myosin rod will

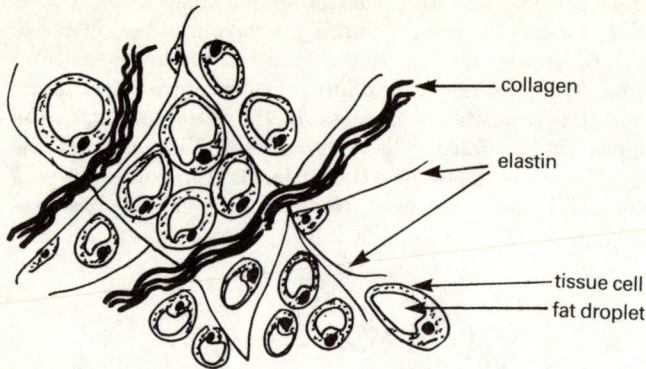

7.6   Structure of fatty tissue

be surrounded by six actin threads set at the corners of a regular hexagon. The actin threads are composed of two strands of sub-units like strings of beads twisted together in the form of a helix. The myosin rods have pairs of short stubby branches, each pair set at 60° to neighbouring pairs so that they radiate from the myosin rods pointing in turn to opposite corners of the hexagon frame of actin threads. A muscle contracts when the sarcomeres contract in unison. When the sarcomere contracts the actin threads run into the bundle of myosin rods and the sarcomere becomes a knot of interdigitated threads. The state of contraction of muscle at the death of the tissue will have profound effects upon the texture of the final product, when the knot of inter-laced actin threads and myosin rods is locked in permanent union in rigor mortis. Indeed, myofibular toughness produced by muscle contraction cannot be rectified by the caterer other than by use of the mincer. The death of the muscle in a good quality carcase occurs some 12 to 18 hours after slaughter. The degree of contraction of muscle and thus meat texture will be affected by events before slaughter which put the animal into a state of stress such as fear, fatigue or exposure to cold, and events after slaughter, particularly chilling or freezing too quickly before the onset of rigor mortis. Thus the caterer should press for humane care in the slaughter of meat animals and the highest technical standards at abattoirs. The quality of meat he puts before his customers largely depends on these factors.

Other organelles normally found in animal cells will also be present. *Mitochondria* are present in large numbers between the myofibrils, particularly in active muscles. *Adenosine triphosphate* synthesized by the mitochondria is the fuel used to power the muscle in life but after death the break-down products of ATP are a principal source of meat flavour. Thus there is a link between activity in life and flavour 'in the pot'. Muscle activity brings an increase in the number of mitochondria and the amount of ATP present in a muscle. After slaughter an increase in ATP residues will provide a richer meaty flavour in the final product.

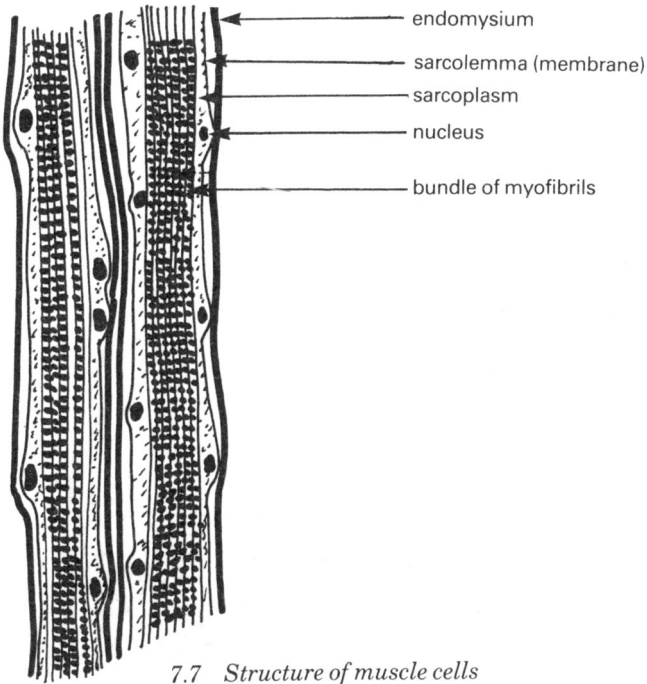

7.7   *Structure of muscle cells*

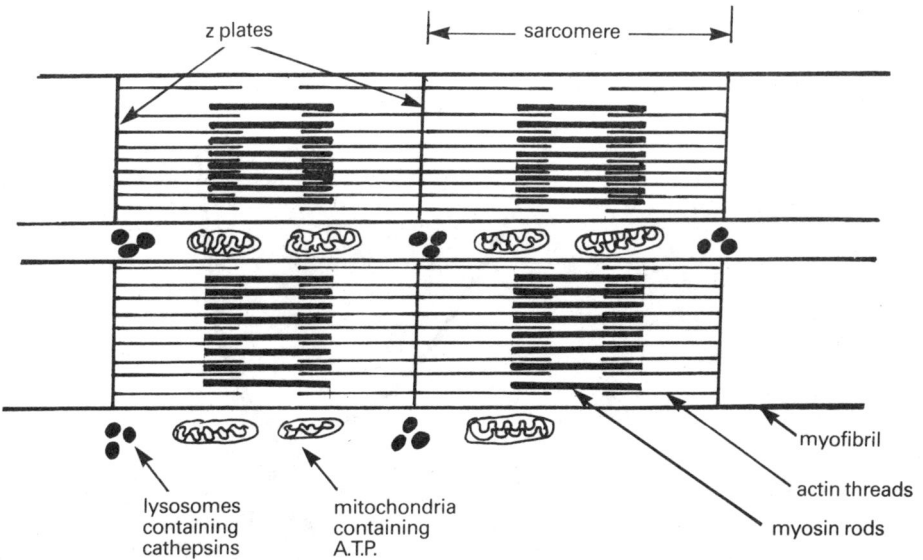

7.8   *Structure of myofibrils*

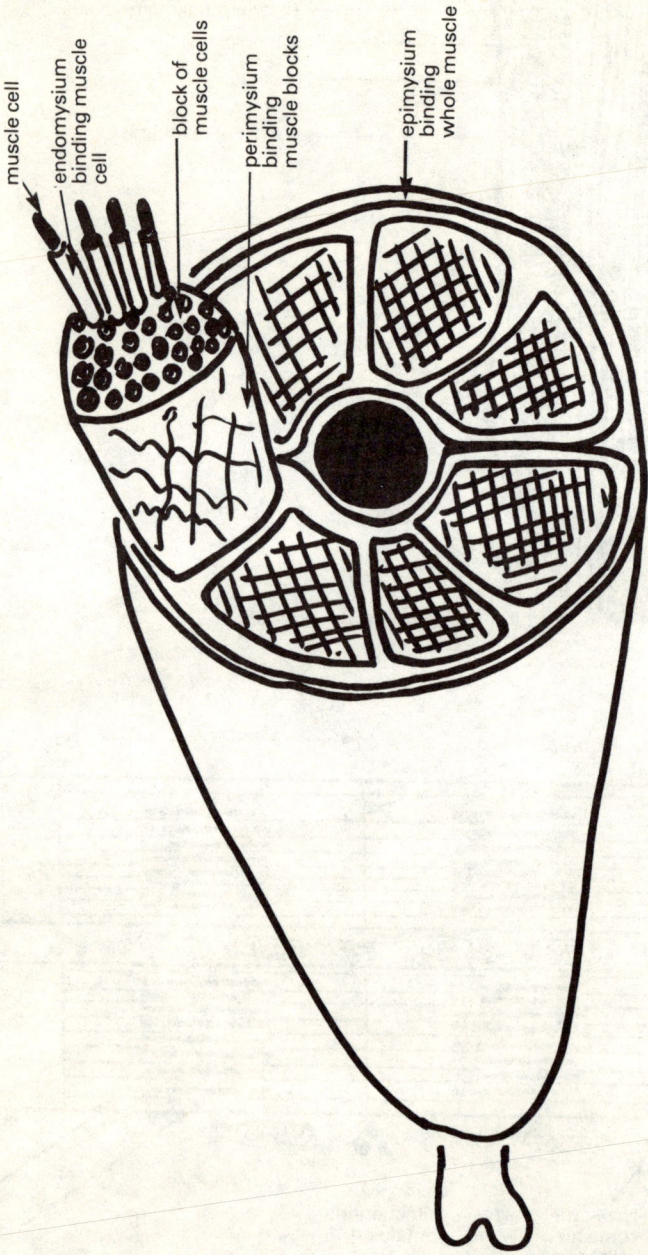

muscle cell

endomysium binding muscle cell

block of muscle cells

perimysium binding muscle blocks

epimysium binding whole muscle

7.9   *Orders of connective tissue*

Similarly between the myofibrils, particularly in the area of Z-plates, are numbers of *lysosomes*. After the death of the muscle these lysosomes release their *cathepsins*, powerful proteases, into the surrounding sarcoplasm. These proteolytic enzymes attack the structures of the cell, in particular the Z-plates, and so bring about softening. The rate of post-rigor softening will depend on carcase temperature. Full softening of beef quarters would require two or three days at 16°C but 10 to 12 days at refrigeration temperatures of 5 to 7°C. The hanging of beef for a week or so allows the process to occur. Sadly commercial pressures to get the meat to market and hold the money in the bank instead, seldom allow for this and both the quality factors of tenderness and flavour are the worse for it.

Muscle as a working unit   Figure 7.9

The muscle is made up of muscle fibres. Each fibre has an outer covering of connective tissue, the *endomysium*. Groups of perhaps a million fibres will be bound into a muscle block by a second layer of connective tissue, the *perimysium*. Then a number of blocks, 5, 10, 20 or more, will be bound into the whole muscle by a third sheet of connective tissue, the *epimysium*. These sheets of connective tissue will extend beyond the ends of the muscle fibres to form the tendons which transfer the power of the muscles exerted in contraction to the limbs.

8

Electricity, Light and Heat

OHM'S LAW

Electricity is a flow of electrons through a wire and, like water in a pipe, the flow requires a difference in pressure between the two ends. In electricity the difference in pressure is measured in *volts* whilst the flow is measured in *amps*. The flow was investigated by Ohm and he found that for a wire held at constant temperature the flow increased proportionally with increase in voltage:

$$\text{Voltage (V)} \propto \text{Amps (I)}$$
$$V = R \times I \text{ (Ohm's Law)}$$

The constant that provides the equation is called the *resistance R* (measured in ohms). The resistance R for a particular material: (a) increases with temperature; (b) increases with length of wire; and (c) decreases with cross-sectional area.

Resistance

All electrical appliances have a resistance and it depends on how these resistances are connected in the circuit what the overall resistance is. If resistances are connected one after each other then they are in series whereas if they are connected at the same points they are in parallel as shown in figure 8.1. The total resistance (R_T) for resistances in series is found by adding the resistances together whereas the R_T for resistances in parallel is found by a more complicated formula:

$$R_T = R_1 + R_2 + R_3 \qquad \text{(series)}$$

$$\frac{1}{R_T} = \frac{1}{R_1} + \frac{1}{R_2} + \frac{1}{R_3} \qquad \text{(parallel)}$$

If we carry out calculations using a fixed voltage of 250v and $R_1 = 25$ ohms, $R_2 = 25$ ohms and $R_3 = 25$ ohms then:

(a) *For series*

$$250 = I \times (R_1 + R_2 + R_3)$$

$$250 = I \times (75)$$

$$I = \frac{250}{75} = 3.3 \text{ amps}$$

(b) *For parallel*

$$\frac{1}{R_T} = \frac{1}{25} + \frac{1}{25} + \frac{1}{25} = \frac{3}{25}$$

$$R_T = \frac{25}{3} = 8.3 \text{ ohms}$$

$$250 = I \times 8.3$$

$$I = 30 \text{ amps}$$

You can see that if 3 resistances of 25 ohms rather than just one of 25 ohms are connected in series then the current is reduced from 10 amps to 3⅓ (to ⅓) whereas if the three resistances are connected in parallel then the current rises to 30 amps such that each of the resistance can have 10 amps (I_1, I_2, I_3 = 10 amps). It is apparent that if you connect appliances in series then the current gets less and less but if you connect in parallel they maintain their own flow but the flow in the main wire increases proportionally. Light bulbs and plugs are connected in parallel with the main wire protected from damage by placing a suitable fuse of 5 amps for lights and 30 amps for plugs in the circuit. Individual appliances are protected by having the appropriate fuse in the plugs.

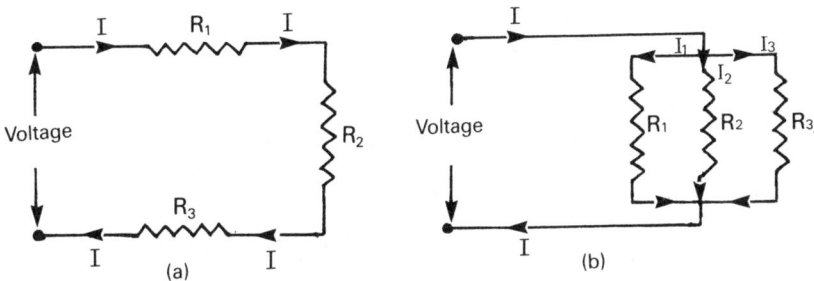

8.1 *Resistances connected in series (a) and paralled (b)*

ENERGY

When an electric current passes through a resistance, energy is released. The quantity of 1 joule of energy is released when a current of 1 amp passes for 1 second under a pressure of 1 volt. If the current or the pressure is increased then the energy per second released is given by the following:

Joules/sec = Volts × Amps

The energy released per second is known as the *electric power* and 1 joule/second is known as a *watt*. Hence:

Watts = Volts × Amps.

This formula can be used to estimate fuse sizes for plugs as all electrical equipment has a wattage given to it. If the mains voltage is taken to be 250 volts then the current required by a 1000 watt appliance (1 kilowatt) is 4 amps and so a 5 amp fuse is required. Likewise a wattage less than 750 watts requires a 3 amp fuse and a 3 kilowatt appliance a 13 amp fuse. Plugs labelled 13 amp is a statement of the maximum current that should be allowed to pass through the plug not a statement of the fuse size it should have.

The unit of consumption of electrical power is the kilowatt-hour, that is a kilowatt on for 1 hour. This corresponds to 1000 watts on for 3600 seconds which results in the consumption of 3.6 million joules. The energy consumption of pieces of equipment of 3 kilowatts or less can be measured using the instrument shown in figure 8.2. The appliance is plugged in to the measuring instrument and a cost per unit of your choice tapped into the display unit. When the instrument is started it records the total cost of your electricity bill which enables you to work out the joules consumed by your appliance using the following formula:

$$\text{Joules consumed} = \frac{\text{Total cost}}{\text{Cost per Unit}} \times 3.6 \text{ million}.$$

The ability to put your own cost per unit into the instrument enables experiments to be carried out over long periods of time (low cost per unit) or short periods of time (high cost per unit). Similar measuring devices can be fitted to larger pieces of equipment using both single and three phase supplies.

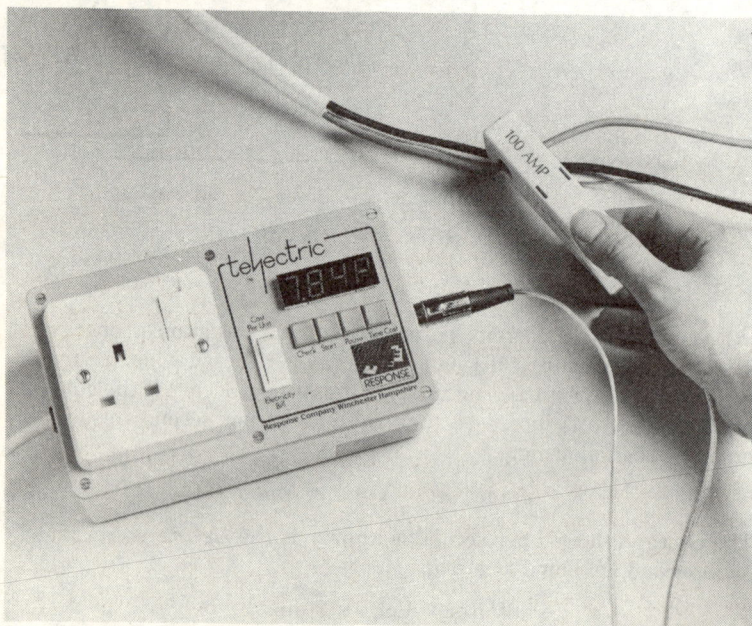

8.2 *Telectric monitor*

The main supply is an alternating current, as shown in figure 8.3, with a frequency of 50 cycles/second (Hz). For small equipment a single phase 250 volt supply is adequate but for large pieces of equipment a 440 volt three phase system is used where the various parts of the equipment take power from each of the various phases as shown in figure 8.4.

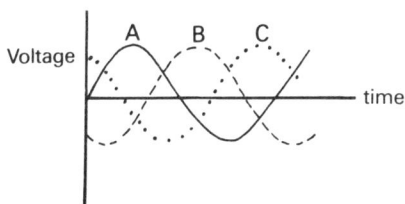

8.3   Single phase (A) and 3 phase power supply (ABC)

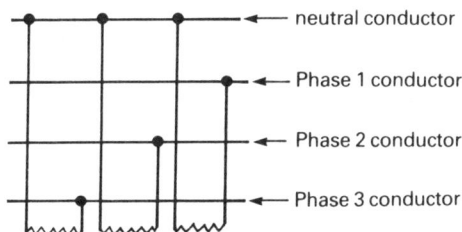

8.4   Equipment using a 3 phase supply

THERMOSTATS AND THERMOCOUPLES

Control over temperature and its measurement are important for cooking purposes. Control is exercised through a thermostat and measurement by employing a thermometer of the thermocouple kind. Thermostats are mainly of two varieties, hydraulic and rod and tube. Both thermostats rely on the expansion of materials on heating to open or close an electrical contact. The important point to remember is that thermostats do not react instantly to temperature changes so there is always a lag causing temperatures to fluctuate around the set temperature.

Thermocouples are constructed by connecting at their ends two dissimilar metals such as iron and copper as in figure 8.5. The difference in temperature between the two ends results in an electrical current that is proportional to this difference. A suitable electronic device placed at the cold end can convert the electrical current into temperature and display it in a digital format. Insulation of the wires is important when using thermocouples in ovens, and metals other than copper and iron are usually employed.

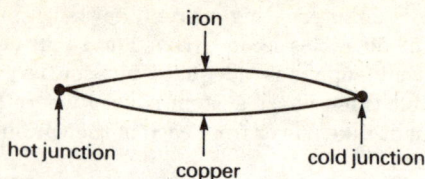

8.5   A thermocouple

INDUCTION COOKING

When a conductor carries an electrical current it produces around itself a magnetic field. If the current is an alternating one then the magnetic field forms, decays and reforms. This changing magnetic field can induce into another conductor placed in that field an electric current (eddy currents) which heat up the metal conductor. This is the principle of the induction cooker (figure 8.6) which produces high frequency power (30 k.Hz) from a low frequency power (50 Hz) to heat up utensils that are made from magnetic materials such as iron and stainless steel. The efficiency of such cookers in comparison with other methods has been investigated (*Journal of Foodservice Systems* 4 (1986) pages 59–66 by Alan Adams and Maurice Palin).

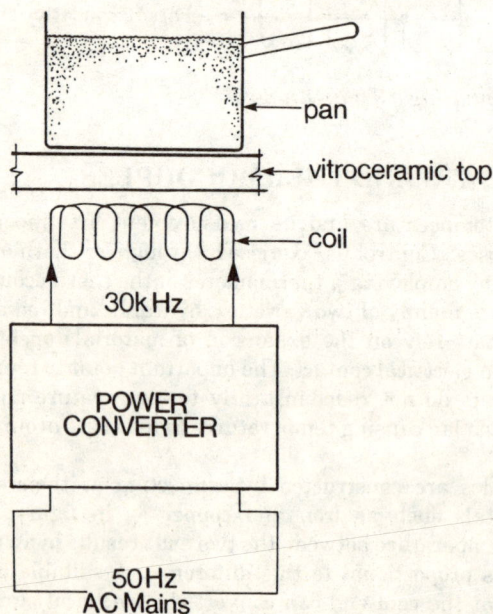

8.6   An induction cooker

LIGHTING

The Food Hygiene Regulations 1980 state that food rooms should be suitably and sufficiently lighted. Suitable lighting is usually white light because under equal lighting conditions the eye, as shown in figure 8.7, is more sensitive to some colours of the visible spectrum than others.

Sufficient lighting is related to the illumination levels under which work is carried out at maximum efficiency. The code issued by the Illuminating Engineering Society gives the following minimum levels of illumination for hotels:

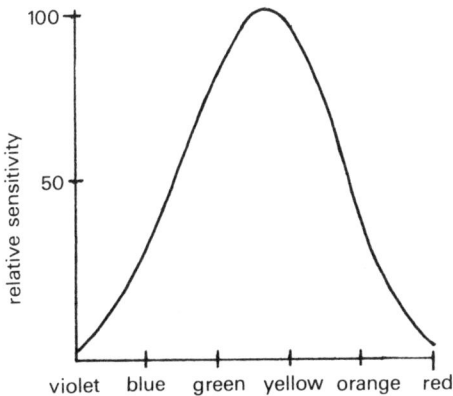

8.7   Relative sensitivity of the eye to the various colours

Table 8.1 *Minimum levels of illumination for hotels*

Area of hotel	Minimum illumination Lux
Entrance hall	200
Reception	400
Dining rooms, grill rooms, restaurants (table lighting extra)	100
Lounges	200
Bedrooms	100
Kitchens	400
Cellars	200

The unit of illumination is the lux and it is the illumination received 1 m away from a standard candle.

Illumination levels fall dramatically the further you go from the light source such that at 2 m the illumination is reduced to ¼ lux and at 3 m to 1/9 lux. This can be expressed by the following formula:

$$\text{Illumination} \propto \frac{1}{\text{distance}^2}$$

The illumination in a particular room can be achieved by putting in the correct number of light bulbs or tubes. The number of bulbs or tubes is best expressed in the total wattage required and this will depend upon:

1   Bulbs or tubes: Tubes are more efficient in producing light than bulbs and so you will require less wattage.
2   Shades will reduce the light and so the wattage requirement will depend on the type of shade used.
3   The height of the ceiling will effect the illumination according to the conditions already discussed.
4   Bulbs and tubes lose their efficiency with time due to deterioration and from the lack of cleaning.
5   The colour of the walls and ceiling will affect the amount of reflected light.

Bulbs produce light by electricity heating up a tungsten filament enclosed in an inert atmosphere, whereas in fluorescent tubes mercury vapour activates a luminescent coating. Bulbs are cheaper, product a warmer light but are less of a supplement to daylight and less effective in distributing their light.

The types of shade that can be used are shown in table 8.2.

Table 8.2   *Types of shade*

Type of shade	Light distribution	Shape
Direct	90% or more downwards	
Semi-direct	60% downwards	
General diffusing	50% downwards	
Semi-indirect	60% upwards	
Indirect	90% or more upwards	

The wattage requirements can be calculated from table 8.3 which relates to a room of (1) height 3 m (2) reflectances of 70% ceiling, 30% wall and 15% floor (3) A = rooms below 12 m^2, B = 12–25 m^2 and C = above 25 m^2.

To calculate the approximate wattage for a kitchen measuring 10 m × 12 m with fluorescent lighting and direct shades we multiply 0.08 × area of the kitchen m^2 × illumination level required in lux = 3840 watts. This involves about 24 double fluorescent fixtures containing two 80 watt tubes. The wattage can be reduced by hanging the light fixtures from the ceiling on chains to reduce the height. A reduction of 1 m can decrease the wattage requirements by less than a half.

Table 8.3   *Approximate wattage requirements*

Fixture and lighting	Approximate wattage per m² to give 1 lux illumination		
	A	B	C
Direct			
Filament (fi)	0.35	0.29	0.25
Fluorescent (fl)	0.12	0.10	0.08
Semi direct (fi)	0.37	0.33	0.27
(fl)	0.12	0.10	0.08
General diffusing (fi)	0.39	0.34	0.28
(fl)	0.13	0.11	0.09
Semi indirect (fi)	0.42	0.36	0.29
(fl)	0.14	0.12	0.10
Indirect (fi)	0.69	0.61	0.50
(fl)	0.29	0.20	0.17

HEAT UNITS

The calorie is the amount of heat needed to heat up 1 g of water through 1°C. The number of calories required to heat up 100 g of water from 20°C to 75°C is $100 \times (75-20) = 5500$ calories. The kilocalorie is 1000 calories and is often used in relation to nutrition where it was until recently called the *Calorie*. The joule(J) is a smaller unit than the calorie and for practical purposes the conversion rate of 4.2 joules per calorie can be used.

The *British Thermal Unit* (Btu) is the amount of heat required to heat up 1 lb of water through 1°F. There is a larger unit called the *therm*, ie 100,000 Btus. Btus. are still used in the gas industry to rate the heating capacity of hot water boilers.

SPECIFIC HEAT

The number of calories needed to raise 1 g of a material through 1°C is not always 1 calorie. This only applies to water, other materials require less than 1 calorie, and the number of calories needed is called the specific heat with the units calories/g/°C. The specific heat of water is therefore 1 cal/g/°C and table 8.4 gives some other specific heats.

Table 8.4   *Specific heats*

Material	Specific heat cals/g/°C
Copper	0.09
Olive oil	0.48
Milk	0.93
Beef	0.77
Beef (frozen)	0.40
Brick	0.22

The number of calories required to heat up 1 kg of beef from 20°C to 39°C is 1000 × 19 ×0.77 = 14630 calories. However the heat required to raise the temperature of 1000 g frozen beef from −20°C to −1°C is 1000 × 19 × 0.4 = 7600 calories.

LATENT HEAT

If you were asked to calculate the calories required to thaw 1000 g of frozen beef by raising its temperature from −20°C to 20°C using the method already discussed you would probably give the following answer:

1   Calories to raise meat from −20°C to −1°C (thawing temperature) = 1000 × 19 × 0.4 = 7600 calories.
2   Calories to raise meat from −1°C to 20°C = 1000 × 21 × 0.77 = 16170.
3   Therefore the total heat required would be 23,770 calories.

Unfortunately you would have not included in your calculation the calories needed to change ice into water (change of state) which occurs at the thawing temperature (−1°C). In other words energy is needed to change state rather than to change the temperature. This energy requirement is called the *latent heat* and has the units calories/g. The latent heat for changing 1 g of ice into 1 g of water at the thawing temperature is 80 cals/g and so if meat contains 75% water then an extra 750 × 80 = 60,000 cals needs to be added to the 23,700 calories which is quite a considerable extra.

The same treatment needs to be applied at the other change of state from a boiling liquid into a gas where in the case of water to steam the latent heat is 540 cals/g.

HEAT TRANSFER

In cooking, heat is usually transferred from a liquid, gas, or solid to the food. The liquid, gas or solid is at a higher temperature than the food and the transfer is carried out by conduction, convection and radiation.

Conduction

Heat transfer by conduction is carried out in a solid where the heat is passed from one particle (stationary) to the next rather like the passing of articles between people forming a chain. In cooking this occurs (1) when food makes direct contact with a hot solid such as a griddle; (2) from skewers placed through food as in kebabs; (3) in solid foods the heat is transferred from the hot surface to the centre by conduction; (4) when heat is lost through oven walls.

The rate of heat transfer under steady state conditions where the temperatures do not change, as in heat transfer from an oven, set a certain temperature to the kitchen is given by the following equation (Fourier):

$$q = kA \frac{t_1 - t_2}{x}$$

where: q = rate of heat transfer (Joules/sec)

A = area (m^2)

t_1 and t_2 = temperatures $°C$

x = thickness of the walls (m)

k = constant called the thermal conductivity (Joules/sec m $°C$).

The equation can be rearranged as follows:

$$q = \frac{t_1 - t_2}{\frac{x}{kA}} = \frac{\text{driving force}}{\text{resistance}}$$

such that the temperature difference $(t_1 - t_2)$ is the driving force for heat conduction and the term $\frac{x}{kA}$ is the resistance to heat flow. The equation looks similar to the flow of electricity where $I = \frac{V}{R}$ (Ohm's Law). It can be seen that the resistance to heat flow is decreased by reducing x but increasing k and A.

The thermal conductivities vary for different materials *such that* metals have high thermal conductivities, whereas air, water and foods have low thermal conductivities. This accounts for the long cooking times of foods using the more traditional methods of cooking.

The equation does not quite satisfy an oven wall because they are usually made of more than one material *such that* the heat has to pass through more than one resistance as in the case of an electric current (figure 8.8).

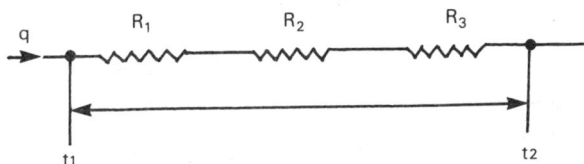

8.8   Resistances to heat flow in an oven wall

The total resistance $(R_T) = R_1 + R_2 + R_3$ and so:

$$q = \frac{t_1 - t_2}{R_1 + R_2 + R_3}$$

The use of the equation can be seen by calculating the heat loss through an oven with the following features: (a) inside temperature 180$°C$; (b) outside temperature 30$°C$; (c) area of 5 m^2; (d) wall consists of an insulating material 20 mm thick sandwiched between two plates of stainless steel each 1 mm thick; (e) the thermal conductivities are 25 m J/m sec $°C$ for the insulating material and 26.4 J/m sec $°C$ for the stainless steel:

$$q = \frac{180-30}{R_1+R_2+R_3}$$

$$R_1 = \frac{x}{kA} = \frac{0.001}{26.4 \times 5} = \text{negligible (outer steel)}$$

$$R_2 = \frac{0.020}{0.025 \times 5} = 0.16 \text{ (insulating material)}$$

$$R_3 = R_1 = \text{negligible}$$

$$R_T = 0.16$$

$$q = \frac{150}{0.16} = 938 \text{ Joules/sec (Watts).}$$

Foods heated by direct contact with a single sided griddle (frying table) have had their complaints because of uneven heating. The temperature over the surface has been reported to vary by as much as 50°C and the power was often found to be insufficient to give reasonable cooking times especially when using frozen products. New methods to improve the temperature distribution over the surface have been achieved using the heat pipe and heat foil techniques. The heat pipe technique (figure 8.9) which involves heat transfer by condensation (latent heat) gives a very narrow temperature spread, low energy consumption, rapid temperature recovery after loading but has a slow start up time. The heat foil is a very thin (0.5 mm) electrical resistance circuit enclosed between electrical insulating films that can withstand temperatures up to 600°C. The circuits are given any design to achieve the desired temperature distribution and are fixed to the bottom of the heating pan. The prototype developed by SIK (Sweden) gave all the advantages of the heat pipe griddle plus a start up time of 4 minutes and incorporated microcomputer control.

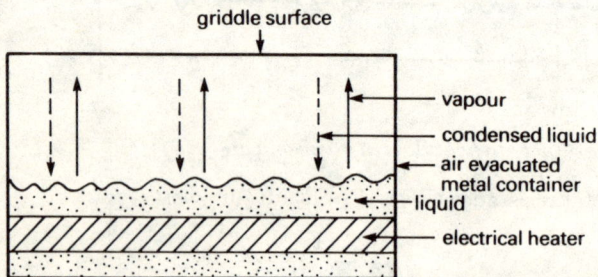

8.9   The heat pipe

Unfortunately the Fourier equation can only be used in calculations where temperatures remain constant. Conduction of heat into foods from direct contact as in the case of the griddle may have the heating surface at constant temperature but the inside of the food is increasing in temperature as the cooking continues. There are however techniques that can be applied in these 'unsteady state' conditions that enable cooking times and food temperatures to be calculated but they are outside the scope of this book (Schmidt's method and the use of published charts by Dalgleish and Ede).

Convection

Heat is transferred to food by liquids and gases (fluids) mainly by convection where the heat is carried by the moving medium in contrast to conduction where the medium is stationary. The use of water, air and fat to carry heat to the food surfaces are well known and so this section will concentrate on the problems of boundary layers which slow down heat transfer.

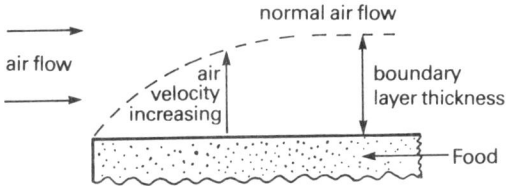

8.10   Boundary layer

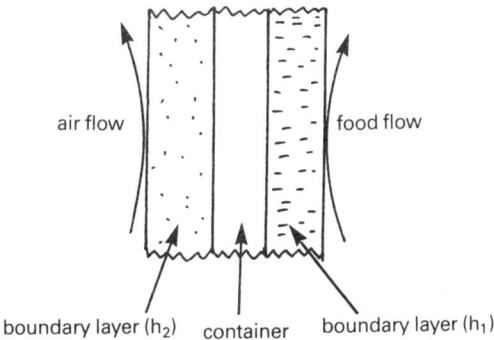

8.11   Boundary layers on either side of a casserole

When liquids and gases flow over a solid surface, as shown in figure 8.10, the fluid is slowed down such that near the surface it is stationary but further away from it the speed increases to that of the normal flow. This makes an extra layer of material through which heat must be conducted and this extra resistance to heat transfer is related to the surface heat transfer coefficient (h). This term can be treated in the same way as any layer of material by stating that $h = \frac{k}{x}$ where k is the thermal conductivity of the boundary layer and x its thickness. The resistance of the boundary layer is therefore $R = \frac{1}{hA}$. When heat is being transferred from air to open food then there is only one boundary layer but when the food is placed inside a container (casserole) then there are two boundary layers and of course the layer of the container in the middle as shown in figure 8.11 (h_1 and h_2 will be different because the thickness and the thermal conductivity of the respective boundary layers will differ). In order to reduce the resistance to heat transfer the boundary layer resistances need to be reduced. The way to do this is to reduce their thickness as we cannot change their thermal conductivities (k). The thickness of the boundary layer on the oven side can be reduced by making the air flow faster over the container which creates turbulent flow when compared to natural convection and reduces the thickness (forced convection ovens). The other method on the oven side is to change the direction of the air as in the jet impinger where the air is directed on to the surface of the food from above and below. These applications are shown in figure 8.12.

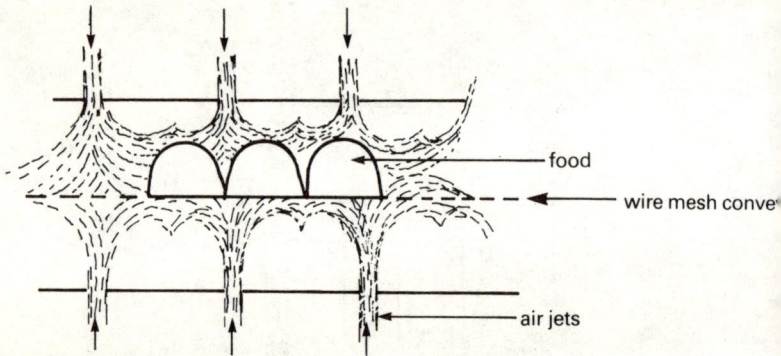

8.12   Jet impinger oven

On the food side, the boundary layers cannot be reduced in such a small container as a casserole dish but in large pans (kettles) the food inside the container can be stirred or in some more expensive pieces of equipment the food can be scraped off the inside practically removing the boundary layer.

Boundary layers also occur in deep fat frying but the expulsion of steam from the food creates turbulent conditions eliminating the need for any mechanical movement of the oil. However the steam itself creates a boundary layer around the food which can be reduced in thickness by putting pressure on the surface of the oil (pressure fryers).

The same layers also occur in water boiling where the heat transfer coefficient to the food increases as the temperature of the water rises due to an increase in natural convection. The low viscosity of water at these high temperatures eliminates the need for a stirrer. Condensing liquids such as steam (steamers and steam-jacketed pans) provide high heat transfer coefficients but a boundary layer of water does form and every effort is made, for example in steam jacket pans, to remove it.

Radiation

Heat transfer by radiation includes infra-red and microwaves. These waves form a part of the electromagnetic spectrum which have different wavelengths but all travel at the speed of light (figure 8.13).

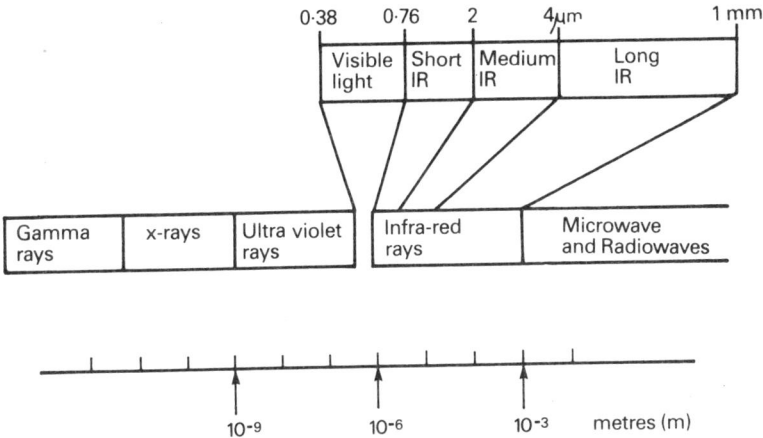

8.13 Electromagnetic spectrum

Infra-red radiation is generated when materials are heated to high temperatures. At temperatures above 2000°C short wave (0.7–2 µm) infra-red is obtained while at lower temperatures either medium (2–4 µm) or long wave (4 µm–1 mm) infra-red radiation is generated. The advantages of infra-red are that it can get straight to the surface of the food (no boundary layers), does not heat up the intervening air and because of the temperatures a high heat transfer rate is possible. The rate at which energy is emitted from a perfect emitter (black body) is given by the following equation which shows the energy to be dependent on the fourth power of its absolute temperature:

$$e = kT^4$$

where e = Watts/m^2

k = Stefan-Boltzman constant

T = temperature of the body in degrees Kelvin (°K)

Infra-red has limited penetration into food, 1.25 mm for short wavelength and 0.5 mm for medium infra-red. The main component of the infra-red oven is the radiator which in the main emits long waves except quartz tubes (medium short) and halogen heaters (ultra-short).

Infra-red ovens have appeared in recent times as regeneration ovens for cook-chill catering systems and in recent publications have been used in continuous ovens, heating from below and above the product. Problems of mechanical damage to pans, soiling of lamps and broken quartz tubes have offset the advantages in reheating and frying catering products. However in infra-red baking the times were reduced by 25%–50%, the energy used was the same, the weight losses were less and the quality as good as the conventional method.

Microwave radiation is obtained from a part of the electromagnetic spectrum with a longer wavelength and therefore a shorter frequency according to the formula c (speed of light) = frequency × wavelength. They are reflected by metals, pass through most plastics and glass and are absorbed by foods. With microwaves it is more usual to refer to frequency rather than wavelength and the two frequencies for microwave cooking are 915 MHz (328 mm) and 2450 MHz (122 mm). All consumer and commercial microwave ovens operate at 2450 MHz of which there are an estimated 50 to 70 million in the use world wide. The principal reasons for choosing these two frequencies are the availability of the microwave generators (magnetron) at these frequencies, the useful power levels (less than 1 kW up to 40–50 kW) and the more efficient heating than at other frequencies. Magnetrons are approximately 50% efficient in converting electricity into microwave power.

Microwaves heat up foods because they create a changing electrical field that causes charged particles to rotate (dipole molecules) or accelerate (ions) thus increasing the kinetic energy of the system which is expressed by an increase in temperature. Foods contain the dipole water molecule and many ions such as Na^+ and Cl^-. Microwave cookery does not involve boundary layers and microwaves score over infra-red in that they can penetrate food to a depth of approximately 2 cm at 0°C but do not cause the browning reactions associated with colour, aroma and flavour. To overcome the latter, browning dishes have been developed that are preheated in the microwave before the food is placed in them or another mode of heat transfer, such as convection, has been incorporated in the oven design.

Microwave ovens, as shown in figure 8.14, are metal boxes which reflect microwaves produced by the magnetron and distributed to the oven cavity by a wave guide. The magnetron needs to be cooled by a fan and the waves are distributed fairly evenly to the food by the use of a metal stirrer and a rotating table.

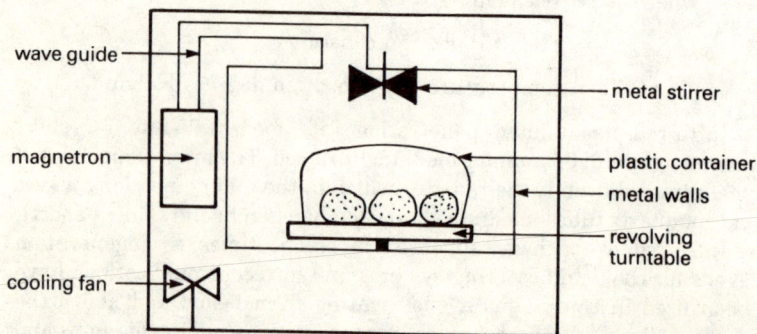

8.14   *Microwave oven*

In catering, microwave ovens of the batch type have found only limited application in areas such as reheating small quantities of cook-chill foods. However, on a larger scale, conveyer belt microwaves have been employed to cook chicken breasts prior to breadcrumbing and freezing using a power of 80 kW and achieving a process time of 14 minutes.

EXERCISES

1 (a) Describe how you would protect electrical wiring.
   (b) Draw the correct wiring diagram for a plug.
   (c) Explain any further safety measures you would take with electrical equipment.
2 Draw the construction and explain the working of any thermostat.
3 Calculate the wattage requirements to adequately illuminate a kitchen of normal height and an area of 25 m^2 with (a) direct fluorescent tubes; (b) semi-direct filament bulbs.
4 What would be (a) the effect on heat loss from the oven already described if all the oven wall thicknesses were doubled; (b) the value of paying for the increased insulation if it cost £50 extra, the oven was operated for 30 hours per week and the cost of energy was 1 p/MJ?
5 Explain how boundary layers form around foods and how the latest equipment has been designed to reduce them?
6 What is the ratio of heat emitted from a radiator at 727°C compared with one at 227°C?
7 Write an account of the latest ovens employing infra-red and microwave radiations outlining their uses in the catering industry.

Index